Military Power and Policy in Asian States

Also of Interest

Huadong: The Story of a Chinese People's Commune, Gordon Bennett

China's Oil Future: A Case of Modest Expectations, Randall W. Hardy

A Theory of Japanese Democracy, Nobutaka Ike

Perspectives on a Changing China: Essays in Honor of Professor C. Martin Wilbur on the Occasion of His Retirement, edited by Joshua A. Fogel and William T. Rowe

Chinese Communist Power and Policy in Xinjiang, 1949-1977, Donald H. McMillen

The People's Republic of China: A Handbook, edited by Harold C. Hinton

Malaysia's Parliamentary System: Representative Politics and Policymaking in a Divided Society, Lloyd D. Musolf and J. Frederick Springer

Mongolia's Culture and Society, Sechin Jagchid and Paul Hyer

The Memoirs of Li Tsung-jen, T. K. Tong and Li Tsung-jen

About the Book and Editors

*Military Power and Policy in
Asian States: China, India, Japan*
edited by Onkar Marwah and Jonathan D. Pollack

This study challenges the belief that the security concerns and strategic objectives of lesser states are dependent on the dominant power alliances and on assessments by major powers of the prospects for peace or war. Focusing on the views of security and military power adopted by elites in China, India, and Japan, the contributors point out that each of these states has been moving steadily toward military autonomy and away from dependence on major powers; while inhibited by the general rules of an international strategic system dominated by the two superpowers, each state has sought to further ends quite different from those assigned them by major powers. This astute description and assessment of the strategies pursued by security elites suggest strongly that future prospects for stability and peace in Asia will be increasingly determined by the states of the region themselves.

Onkar Marwah is assistant director, Program for Strategic and International Security Studies, Graduate Institute of International Studies, University of Geneva, and served for seven years with the government of India's Administrative Service. Jonathan Pollack is a member of the Social Science Department of the Rand Corporation.

Military Power and Policy in Asian States: China, India, Japan

edited by Onkar Marwah and Jonathan D. Pollack

Contributors:
Onkar Marwah
Jonathan D. Pollack
Stephen P. Cohen
Yasuhisa Nakada

Westview Press • Boulder, Colorado
Dawson • Folkestone, England

This volume is included in Westview's Special Studies on China and East Asia/South and Southeast Asia.

All rights reserved. No part of this publication may be reproduced or transmitted in any form or by any means, electronic or mechanical, including photocopy, recording, or any information storage and retrieval system, without permission in writing from the publisher.

Copyright © 1980 by Westview Press, Inc.

Published in 1980 in the United States of America by
  Westview Press, Inc.
  5500 Central Avenue
  Boulder, Colorado 80301
  Frederick A. Praeger, Publisher

Published in 1980 in Great Britain by
  Wm. Dawson and Sons, Ltd.
  Cannon House
  Folkestone
  Kent CT19 5EE England

Library of Congress Catalog Card Number: 79-16241
ISBN (U.S.): 0-89158-407-2
ISBN (U.K.): 0-7129-0890-0

Printed and bound in the United States of America

To our mothers
Smt. Trilochan Kaur Marwah
and
Esther Duker-Pollack

Contents

List of Tables xi
Preface, *Onkar Marwah and Jonathan D. Pollack* xiii
The Contributors xvii

1. Introduction: Asia and the
   International Strategic System,
   Onkar Marwah and Jonathan D. Pollack 1

2. Toward a Great State in Asia? *Stephen P. Cohen* 9

3. China as a Military Power, *Jonathan D. Pollack* 43

4. India's Military Power and Policy,
   Onkar Marwah 101

5. Japan's Security Perceptions and Military Needs,
   Yasuhisa Nakada 147

Tables

2.1 Defense Spending, Population, and Economy 15

2.2 Force Levels 16

2.3 Relative Ranks 17

2.4 Relative Burden of Military Expenditures 18

4.1 Ability to Manufacture Weapons:
China and India, 1979 (Selected Items) 120

4.2 India: Defense Expenditures as a
Percentage of GNP, 1950-1978 129

4.3 India: Foreign Arms Purchases, 1965-1975 132

5.1 Japan's Defense Expenditures and the
Strength of the Self-Defense Forces 173

Preface

The need to assemble a volume on the security perspectives of Asia's major states has been apparent to both of the editors for some time. For far too long, China, India, and Japan have been viewed as states with regionally derived national security objectives and policies. In a superpower context, this judgment is no doubt correct. However, reliance on this single perspective is both limiting and shortsighted, as witness the dramatic events within Iran and between Vietnam and Cambodia without deference to superpower concerns. Indeed, the Asia of the 1970s is most assuredly not the Asia of the previous decades. This conclusion will apply with far greater force over the next ten years. A serious consideration of the independent directions and consequences of the political and military changes now underway among Asian states is thus long overdue.

However, simply to assert that Asia is changing tells us little. What we have tried to do in this collection of essays is to look at the more concrete meaning and implications of this transition and at where it is likely to lead in the coming decade. The fact that it is now possible to assemble this collection is indeed reflective of the heightened consciousness about security and military power in these societies. Such developments are unmistakably underway, and not to take stock of them seems myopic at best.

In preparing this volume, we have made no attempt to impose a coherent overall view. The differences in approach among the various authors should be readily apparent. However, all of the contributors share an overall perspective concerning the study of military power and policy. We have therefore sought whenever possible to have a common set of issues discussed by each author. Some of these broader questions are further considered in our introductory chapter and in Stephen Cohen's essay.

The immediate stimulus for this volume came from a panel we jointly organized for the eighteenth annual convention of the International Studies Association, held in St. Louis in March 1977. On the suggestion of Mervyn Adams Seldon, we recast the initial papers and presentations, and in some cases sought new contributors. Though this process has led to a delay in the volume's publication, we believe that a far stronger set of essays has been the result. Lynne Rienner of Westview Press deserves special mention for her reminders (at first gentle, but somewhat later justifiably more blunt) that delinquent editors necessarily deserve. Her forbearance, always with an appropriate measure of humor and friendliness, surpassed all reasonable limits.

We also wish to acknowledge the generous financial support without which this volume would never have appeared. The Harvard University Program (now Center) for Science and International Affairs provided a supportive intellectual environment where such collaborative work was actively encouraged and facilitated. In addition, the facilities of the Fairbank Center for East Asian Research were of considerable help to Jonathan Pollack. During the 1977-1978 academic year, both editors received fellowships from the Program in International Relations of the Rockefeller Foundation. The foundation's generous support enabled us to continue our research and complete work which would otherwise have languished.

Our dedication reflects our parallel indebtedness to our mothers. Their influence is evident not so much in our actual research, but in the personal and intellectual encouragement they have long provided. In a more immediate sense, Jonathan Pollack wants to thank both Josh and Noah, who (even if they do not yet understand publishers' deadlines) surely sensed when work had to intrude on our time together.

With the exception of minor revisions and additions, this volume was completed before Jonathan Pollack joined the research staff of the Rand Corporation. The opinions and judgments in his contributions to this book are in any case entirely his own, and do not reflect the views of Rand or any of its government sponsors.

Onkar Marwah
Jonathan D. Pollack

The Contributors

Stephen P. Cohen is associate professor of political science and Asian studies at the University of Illinois. He has written extensively on military systems in South Asia, on United States policy towards India and Pakistan, and is presently engaged in a study of regional arms control and proliferation issues. He is author or coauthor of several books, including *The Indian Army* and *India: Emergent Power?*

Onkar Marwah is currently assistant director, Programs for Strategic and International Security Studies, Graduate Institute of International Studies, University of Geneva. He holds a Ph.D. in political science from the University of California at Berkeley, as well as degrees in economics from the University of Calcutta, the London School of Economics, and Yale University. Dr. Marwah has taught international relations at Clark University, has been a visiting associate professor of economics in the Pace Program of Old Dominion University, and held a research appointment with the Peace Studies Program of Cornell University. Prior to his academic career, Dr. Marwah was employed by the Indian government as a member of the Indian Administrative Service. He has contributed to a number of journals on Third World security concerns, coedited *Nuclear Proliferation and the Near Nuclear Countries*, and published *Growth and Modernization in India and China*. He is completing a study of India's military system and is also at work on a research project on nuclear proliferation among six developing countries as a fellow in the Rockefeller Foundation Program in International Relations.

Yasuhisa Nakada, journalist with the political section of Kyodo News Agency, Tokyo, currently covers Japan's Defense Agency and the Ministry of Foreign Affairs. He graduated from Waseda University in Tokyo and has contributed a number of articles on Japan's diplomacy and security issues to monthly magazines in Japan. During 1977-1978 he received a grant under which he pursued arms control research at Harvard University.

Jonathan D. Pollack is a staff member of the Social Science Department of the Rand Corporation. Prior to joining Rand he held research appointments at Harvard University in the Program (now Center) for Science and International Affairs and the Fairbank Center for East Asian Research. He has also taught at Brandeis University. A specialist on Chinese foreign and security policy and the international politics of East Asia, he holds M.A. and Ph.D. degrees from the University of Michigan. He has published research in various journals and in edited volumes, including *Great Issues of International Politics, Political-Military Systems: Comparative Perspectives, Asia's Nuclear Future, The Soviet Threat—Myths and Realities,* and *The International Political Effects of the Spread of Nuclear Weapons.* He is also revising his doctoral dissertation, *Perception and Action in Chinese Foreign Policy,* for publication as a book.

Military Power and Policy in Asian States

1
Introduction: Asia and the International Strategic System

Onkar Marwah
Jonathan D. Pollack

During the past decade, both scholars and practitioners in the field of international politics have experienced a growing unease about the validity and viability of long-prevailing political and intellectual beliefs. The predominant focus on the European state system in international relations theory and practice has seemed increasingly questionable in light of the decline of the major colonial regimes and the concomitant emergence of the United States and the Soviet Union as the world's preeminent military powers. The singular importance of postwar U.S.-Soviet relations and the strategic nuclear competition that so shaped these ties testifies to the existence of a vastly different structure of international power. Yet, despite the change, scholarly effort has failed to produce a newer and more coherent conception of the international state system and the institutional and legal norms underlying it. As a result, past assumptions—in particular, those ascribing to the European state system universalistic norms about the exercise and management of power—have persisted.

The continued erosion of European power has been an inescapable trend over the past thirty years. Further, the global rivalry between the superpowers has frequently obscured a broader process of international change that has been increasingly evident during the past decade. It is this latter process—a widespread diffusion of political, economic, and military power—that the essays in this volume attempt to address. Though this process has been long underway, scholars have been somewhat slow in reacting to such change and its implications. Comparative political studies since World War II have been oriented principally toward the internal dimension of political change, with particular attention to issues of political participation, institutionalization, and economic modernization. Without denigrating this prodigious and frequently impressive research effort, the categories employed in such scholarship can still be considered unduly restrictive. The inattention to the external realm—both as a constraint upon and stimulant to political action as well as the object of much elite activity—is all too apparent. The study of military behavior and institutions has also been almost exclusively within the context of the domestic politics of various regimes.

Yet newly established regimes in Asia have not developed in a vacuum. They have emerged within the context of an unstable (and frequently threatening) international environment—with external imperatives frequently encountered in terms of a superior military presence. The vulnerabilities induced by the technological, material, and organizational power of the West both in the past and present have been a source of frustration for indigenous elites. What Chinese have termed their "century of shame and humiliation" applies elsewhere in Asia with comparable force and conviction. The effort to rectify these disparities in national power has been a central element in the brief histories of states throughout the Asian world.

In this volume, attention is focused on three Asian societies—one already industrialized and two in the midst of modernization—that have sought to come to grips with such questions. In considering these three to the exclusion of many others, we are vulnerable to the charge of a major-power fixation. Yet the diffusion of power is a vastly understudied topic. Inquiry must therefore begin with those Asian states most acutely conscious of the external world and their relationships to it.

An additional question raised by our orientation concerns its focus on military power and policy. There is a widespread belief that it is unwise and indeed illegitimate for states, especially in the Third World, to expend considerable energies and effort on acquiring the instruments of violence. Three principal reasons underlie these objections. First, it is feared that a use for such coercive power is more likely to be found in a domestic context than in the maintenance of territorial integrity and national sovereignty against external threats. Second, it is widely assumed that attention to the military realm can only be to the detriment of extremely pressing social, economic, and educational inequalities. Finally, the very objective of acquiring and employing force in relation to external goals is widely questioned.

However, as some of the authors have indicated in their essays, the validity of these arguments—in particular the second, which deems defense and development separable and antagonistic objectives—is open to some question. Moreover, these critiques ultimately reflect divergent political values on the part of those either articulating or challenging the worth of military power. This essay will not attempt to resolve these difficult issues in any conclusive way. Much of this volume, however, seeks to come to grips more fully with why and how the accumulation of military capabilities has been deemed a vital need in the Third World. Any arguments seeking to deny the validity of acquiring such power must somehow override an incontrovertible fact—an acute remembrance within these societies of their vulnerabilities when superior force was employed against them.

Thus, whatever policies now exist or are likely to emerge must be seen in terms of the inescapable overlay of the past. In various ways, each of these states has had to operate within the context of technological, economic, and diplomatic constraints that have impinged substantially on their freedom of action. The remembrance of such constraints occupies a central position in the political and military conceptions that elites in Asia and elsewhere have adopted.

We do not consider past dependencies as immutable or permanent. Indeed, the recent history of each state considered in this volume reveals a progressive growth in national identities, with indigenous elites increasingly able to chart their own course vis-à-vis the outside world. In this respect, the realm of national security is far from unique. However, it is central in different ways to their respective experiences, and continues to vitally affect each of these polities, albeit on varying terms. The future relationship between the major states of Asia and a global strategic system will continue to reflect such historical remembrances; this factor should be kept in mind throughout these essays.

Regardless of the significant accumulation of power in these societies, the perception of this power and the leverage it affords them remains obscured. There is a tendency in scholarship as well as politics to view China, India, and Japan as essentially reactive in their power and policy. A related premise is that "central system" actors can still effectively curb their exercise of power. No doubt the external capabilities of all three states (in a purely military sense) remain highly constrained. Yet it does not follow that they can thus be readily assimilated within various centrally conceived and managed security systems. While each of these actors has accommodated to a larger international security system, and will continue to do so, the process is no longer one of wholesale acquiescence to great power norms. It rather represents a conscious strategy in two senses: first, to greatly raise the stakes of any external actor seeking to exercise military power with impunity against them; and second, to develop capabilities that address the real needs of national security for these states in the context of their own regional environments.

An appreciation of both facets of military power in Asia remains lacking in most efforts to conceptualize the security needs of noncentral system actors. Arguments about which states are most appropriately included among the "regional influentials" obscure the ongoing efforts of these states as well as others to rectify acutely felt disparities and vulnerabilities. The consciousness and determination with which these efforts are being pursued in Asia are too often ignored or disparaged. Over time, new security arrangements in East, South, and West Asia have slowly but inescapably started to emerge, as the abrupt changes within Iran indicate. The specific configurations of power remain very much in flux. Thus, there is not much merit at present in undertaking highly detailed projections on the prospects for peace, war, or external involvement. What these essays do attempt to sketch is when, where, and how elites in these societies (either in the past or at present) have seen their security as threatened, and what kinds of efforts have been undertaken to deal with such conditions.

What is the role of military power in effecting the transition to more autonomous regional security arrangements? Classically, the instrument of armed force and the threatened or actual use of violence has been justified as a means of achieving greater control and predictability in the international system. Though there is a self-serving aspect to such a rationale, this function is still very much evident. The ability of leaders to influence (if not wholly determine) the way in which their nation will interact with a larger international system remains highly coveted, but very difficult to achieve. This is a goal which all three states considered in this volume have consciously sought. Though the power of external actors remains resilient and is obviously still very substantial, the "rule-changing" capacities of these states in relation to Asia's emergent powers have been progressively reduced. It would be presumptuous to try to predict with any degree of certainty the ultimate result of this transformation, and we will not attempt to do so. However, we should remain sensitive to the circumstances, conditions, and events which testify to such change.

Regardless of the political, ideological, and economic differences among these states, all have sought to change from a subjected position to one that is less physically vulnerable, and hence more secure. We are looking at the military dimension not because it is the exclusive focus of this effort, but because it constitutes a particularly revealing means by which these three states have sought to cope with the outside world. There are, to be sure, inescapable dilemmas and ironies in such reliance on military power. For all three, past military deficiencies were a principal cause of the outside world successfully encroaching upon them. For Japan, the possession of substantial arms became an instrument of expansion abroad. Thus, the purposes to which such capabilities have been put can be a source of restraint as well as an incentive to acquire military power. At the same time, there is obviously no simple relationship between the accumulation of such power and a corresponding increase in political leverage or influence. Two considerations, therefore, equally merit attention: the potentialities inherent in each of these states in the realm of modern military power; and the political, diplomatic, and strategic purposes toward which such arms might be directed.

Analyzed in a static sense, the absolute quotients of power presently available to Asia's major powers remain modest by comparison with the superpowers or indeed some of the European powers. Such comparisons, however, reflect past constraints (political, technological, and economic) affecting the participation of these societies in the competition to acquire modern arms. As the individual chapters indicate, with the exception of Japan, Asian states have embarked upon their efforts in a sustained way only in the relatively recent past. Moreover, the appropriate comparison is less with the capabilities of external powers, but more in relation to what the various Asian regions and security environments might look like in the absence of indigenous military strength. Though each of these states in various ways is conscious of its broader international role, all presently remain regional rather than global powers. By basing our assessment on the latter criterion, a more meaningful appreciation of the significance of their military efforts is possible.

Irrespective of any political or ideological quirks that might occur within any of these polities, the weapons development and acquisition process is now a sustainable national objective in all three regimes. While this does not guarantee predominance by any of them within their own regional environments, it greatly reduces the possibility that they will be subject to severe external compulsions.

Equally important, these states have a political conception of their power which their emergent defense capacities can be expected to serve. While all three have fostered or maintained considerable external involvements and political commitments, an underlying adherence to "nonalignment" is also discernible. Our use of this term is not in its rhetorical sense or as a reflection of the stark bipolarity of the early postwar world. Rather, pursuit of nonalignment is indicative of a widespread urge among such states to define in independent, active terms the level, form, and extent of their participation in the external world.

Introduction

In this respect, China, India, and Japan are all challenger-states, irrespective of whether they call themselves aligned or nonaligned, revolutionary or status quo. If not an immutable or irreversible trend, it is the predominant direction for the politics and strategy of each. Regardless of the extent to which any of them have in the past been protected by an external great power—or made over in the image of a central-system state—a radically different process and outcome are now in motion. In military terms, all three societies are actual or nascent powers within their own regions. In political-diplomatic terms, this urge for autonomy is even further advanced, and will not be lightly or easily reversed. It is an understanding of this dual process of power acquisition—developing both the attributes and ideas underlying political, diplomatic, and strategic independence—that students of international security must now seek.

Thus, if the field of international relations requires new conceptualizations to replace outworn ones, the issues raised in this volume merit consideration. Bipolar conceptions of world politics increasingly seem as overdrawn as the Eurocentric views that preceded them. Regional powers such as China, India, and Japan are not only beginning to articulate a role of greater independence and national initiative; they are successfully implementing such policies. Moreover, past discrepancies between the conception of power and the actual capability have been reduced. If independence is assessed in terms of the capacity to resist external pressure as well as convey and carry out indigenous conceptions of security, then each of these states has made significant progress toward achieving this objective. Wherever international politics may be headed, lesser rather than greater concentration of power seems certain. Such a process will surely require far greater effort to assess the political and strategic consequences of a more complex, differentiated world. The individual essays in this volume hopefully constitute a modest first step in that direction.

2
Toward a Great State in Asia?

Stephen P. Cohen

States strong enough to do good are but few.
Their number would seem limited to three.
Good is a thing that they, the great, can do,
But puny little states can only be.

—Robert Frost, "The Planners," 1946

Introduction

In Bismarckian Europe the presence of a great power could be readily determined. It was a state that was able to decide on its own whether or not it would engage in warfare.[1] The remaining nations were subordinated to the will or whim of the great states of Europe. With the qualified exception of imperial Japan after 1905, the suggestion that great powers existed outside of Europe was laughable. Retrospectively, the widespread notion that only within such a great power could the individual achieve true self-fulfillment is of equal interest. The greatness of the state was believed to enhance the quality of citizenship, a conviction that dates from as early as the Greek city-state.

One hundred years later such views are simply not uttered in polite society. All states—great and small—are deemed equal, at least before the UN secretariat. Universal democratic egalitarianism encompasses relations between most states (if not within them) and all, large and small, are equally great powers—a nonsense which may be marginally functional in taming the excessive pride and ambition of some of their numbers.

The determination of great-power status has not advanced much beyond Treitschke's cynicism. Outside of the two superpowers no state can apply its military power at will and there are serious (and increasing) limits on even their capacity to intervene.[2] Some would ask of what use is military power in an era of nuclear stalemate, exploding populations, fragile economies, and disintegrating societies? How can Third World states, in particular, rationally (let alone morally) contemplate the acquisition of enhanced military power when such power will probably be used by the military against the government in power or the population in the streets?[3] In sum, great powers as we have known them—as military powers—are said to be obsolete, if not more dangerous to themselves than to others.

Some, however, still maintain the old arguments about the necessity of force and power. Morton Kaplan has recently tried to revive Mackinder's geopolitics, and there has been a curious effort to quantify various aspects of "perceived power" by Ray S. Cline.[4] Both Kaplan and Cline argue for the existence of critical regions surrounding the global heartland, and Cline suggests a new "Athenian Alliance" to the United States and the Western European powers, who must associate themselves with at least one great power in each major "politectonic" region.[5]

Without completely rejecting the utility of such approaches to the analysis of great powers, this volume grows out of quite a different perspective and should not be confused with this new school of regional power politics. Between the superpower and the smaller (or fragile) state there are perhaps a dozen or so middlepowers.[6] At least three of these middlepowers are in Asia and appear to be increasing their relative advantage over both immediate neighbors and distant states. Collectively, the chapters in this book raise the question of the emergence of one or more of these states—China, India, Japan—as a latter-day version of Europe's great powers. Even while accounting for the differences in the international system in which they function, the presence of two dominating superpowers, their internal weaknesses, and the uncertain impact of military technology, they will ultimately be regarded as qualitatively different from the next order of states. In Asia these states include Indonesia, Iran, Pakistan, and Vietnam. We do not make the collective claim that China, India, and Japan form the core of their regional systems, although some analysts do and others still may. If we have a shared concern, it is that the influence and power of these states be neither overestimated nor underestimated, hopefully avoiding such grotesque estimates as Cline's ranking Pakistan ahead of India, the hysteria surrounding the construction of an "anti-Chinese" ABM in the United States, the still-heard belief that Japan could and should be pushed into the world as an independent power center, or the exaggerated hopes and fears surrounding the future of Iran.

What does the term "great power" mean in contemporary Asia? At the very minimum, it implies regional hegemony or rough equality with a neighboring great power. It may further include continental or global influence—although not necessarily dominance. The attainment of such power can come about only if a number of conditions are fulfilled. Most, but not all, pertain to the enhancement of national capacities:

- The capacity to manage the domestic processes of economic development and national integration
- The capacity to resist outside penetration
- The capacity to dominate regional competitors
- The capacity to deter outside states (especially superpowers) from lending support to regional competitors
- The capacity to achieve autarky in critical weapons systems, or at least to be able to bargain successfully for them during crises
- An awareness that the above capacities exist for—or are within the reach of—a state increasing its relative strength and influence

Great power status thus implies the existence of local military preponderance over neighbors through the spectrum of force and the means to maintain that dominance. It may also include the availability of nonmilitary instruments of statecraft, the ability to manipulate the domestic political weaknesses of rival states, and certainly a willingness to conduct a diplomacy that places power and status over other objectives. Finally, if necessary, a great power is willing and able to make external political commitments and has the resources to fulfill such commitments. We do not refer to the commitment of the weak alliance partner to the strong, but of the strong to an equal or a weaker state.

Do China, India, and Japan fulfill these criteria? It is an assumption of the collective effort in this book that they do, at least in part. The individual-country chapters will probe the accomplishments and deficiencies of each state in greater detail in terms of its great power potential, but we can first offer some general and comparative observations. These observations are of two kinds. The first pertain to the general and shared characteristics of China, India, and Japan as legatees of diverse Asian traditions, which also possess some of the material and political requisites for great power status. The second type of observation in our analysis will attempt to go beyond the individual-country chapters, which are largely straight-line projections of recent historical trends, and identify some areas of uncertainty and instability in the prediction process. These include the influence of technology on military balances, trends in the nuclear proliferation and arms control process, and the perception and behavior of the great powers.

Three Great Powers—or None

On Having a History

The meaning of history is a subjective matter, determined as much by present reflection on the past as by the actual course of events. As Adda Bozeman has written:

> What once was actual and real often loses its poignancy for later generations, and what is only dimly discernible in the present may assume clear contours sometime in the future.... Memories that control the thinking of the Arabs may not suggest significant recollections to the Chinese or the French. Each nation, then, has a different image of the past, and each generation a different perspective over the passage of time.[7]

The fact that the three states we are examining are Asian is important both because of certain historical traditions and because of their self-perceptions and the perceptions of outside states towards them.

The relevant historical legacy of China, India, and most recently, Japan, is that they were once triumphant states or empires with a continental reach. This is as critical to their emergence as great powers as any "Asian-ness" that might exist. Each of these states has a great tradition of imperial or continental power, and in each case this tradition stands in sharp contrast to more recent subservience, humiliation, and vulnerability.

The vulnerability of Asia to the West (and much of Latin America and Africa, as well) was two-fold. First, Europe and Asia developed only the most rudimentary common diplomatic language. From their earliest contacts there were ready exchanges of technology and material goods, but there was no growth of a shared view of an ordered world system. Second, in the absence of such a shared view, power and force became the ultimate arbiter of differences and the dialogue between the Orient and the Occident was essentially a "dialogue des sourds" (dialogue of the deaf).[8] In this dialogue the various Asian powers were extremely vulnerable to the numerically inferior but highly motivated and very well-organized military forces of Portugal, France, Great Britain, Holland, Germany, the United States, and Russia. Through bitter experience it has become clear to the elites of these Asian nations that their military vulnerability was the critical factor explaining their weakness, not any inherent failure of their society or culture. While compromises had to be made to accommodate Western military technology (including major changes in their societies) this did not imply the destruction of indigenous values or the abandonment of idealized models of the past.

In China, India, and Japan, military power and technology became an obsession as a result of contact with the West, and the individual-country chapters necessarily focus on this issue. To the elites of these states (with the special exception of Japan, which had passed through this phase by 1945), the liberation of their countries from Western influence or dominance, the reestablishment of former imperial or hegemonial relations, as well as the establishment of control over their own societies even today require substantial military power. Ironically, such power was and still is often derived from the West, leading to a troubled relationship between these states and potential Western arms suppliers. The recent histories of these states, unlike that of a number of other Afro-Asian states, have as much to do with ambivalent military relations with the West as with economic dependencies. For if a state has even a token interest in restoring something of its historical position, it must inevitably treat economic power as a means to military power, and use the latter in the service of a diplomacy whose goal is the shaping of a desired regional or world order.

Material Basis of Power

If Western Europe and the United States were the proper models for great power or superpower status, then only Japan would approach the standard, and then with some difficulty. India and China are not rich or developed in the Western sense. Yet, because military power is relative and takes several forms, these states do begin to fulfill some of the material requirements for great power status.[9] While their economic systems differ greatly, as do their populations (in terms of education, skills, health, and social organization), they do share one characteristic: they are able to generate large surplus funds, manpower, and/or advanced technology to enable them to support a substantial military effort. Japan alone has not done so and spends under 1 percent of its GNP on defense, but it has the material capacity to develop into a great military power in the future. The inhibitions which limit defense spending in Japan are political and psychological, not economic, which does not imply that Japan will necessarily increase its defense burden.

The accompanying tables indicate both the enormous magnitude of forces and the defense effort of which a number of Asian states are capable, and points up their structural differences. Only India and China are similar in their economic base: large, poor peasant societies with relatively small (but in absolute terms, substantial) industrial sectors. The enormous populations can be organized into huge

TABLE 2.1 Defense Spending, Population, and Economy

	Population (in millions)	GNP (billion U.S. $)	MILEX* (billion U.S. $)	Armed Forces Manpower (000)	Percent GNP Spent on Defense	MILEX Pop.	MILEX Manpower	No. Civilians per soldier	% men 18-45 in millions
China	925	$299	$25	3,950	7.7%	$26	$6,300	237	2.2%
India	622	$89.7	$3.45	1,096	3.8%	$5.5	$3,147	567	8%
Iran	35	$56.8 (est. GDP)	$7.9	342	14% (GDP/MILEX)	$228	$23,000	101	5.1%
Japan	114	$567	$6.1	238	1%	$54	$25,600	478	9%
Indonesia	135	$29.2	$1.35	247	4.6%	$10	$5,465	546	1%
Pakistan	74	$10.1	$.819	428	8%	$11	$1,913	173	4%

Source: All data are for 1975-1976, and are derived from the International Institute for Strategic Studies (IISS), The Military Balance 1977-1978 (London: IISS, 1978). Certain figures are rough estimates (China's population, economy); others are not exactly comparable because of different measuring techniques.

*MILEX--military expenditures

TABLE 2.2 Force Levels

	No. Equivalent Infantry Divisions	No. Armored Division Equivalents	No. Heavy, Med. Tanks	No. Combat Aircraft	No. Submarines	No. Major Surface Vessels
China	130**	12	8,000 (est)	5,200	67	22
India	27	4	1,780	700	8	31
Iran	5	3	1,620	350	(3 on order)	11 (6 on order)
Japan	12	1	610	456	15	45
Indonesia	4	.3	---	30	3	11
Pakistan	15	3	1,000	247	3	5

Source: Derived from individual country entries in the International Institute for Strategic Studies (IISS), *The Military Balance 1977-1978* (London: IISS, 1978). Light tanks, armored personnel carriers (APCs) are excluded, as are patrol vessels. There is a great variation in modernity and readiness of equipment.

**Excludes est. 100 local force divisions.

TABLE 2.3 Relative Ranks

	MILEX* Per Capita	MILEX Per Soldier	Population Per Soldier	Literacy	Life Expectancy	Calories Per Capita
China	$60	$53	69	24	53	97
India	$95	$89	95	83	81	112
Iran	$22	$22	43	80	81	80
Japan	$42	$20	83	1	5	56
Indonesia	$105	$109	82	65	89	129
Pakistan	$80	$111	46	107	82	100

Source: Derived from Ruth L. Sivard, ed., World Military and Social Expenditures, 1976 (WMSE Publications: Leesburg, Virginia, 1976). Based on 1973 data from 132 countries; rankings are approximations.

*MILEX--military expenditures

TABLE 2.4 Relative Burden of Military Expenditures*

Military Expenditure as Percent of GNP	Per Capita GNP						
	Less than $100	$100-199	$200-299	$300-499	$500-999	$1,000-1,999	above $2,000
more than 10%	North Vietnam		Egypt	North Korea Jordan Syria	<u>Iran</u> <u>Iraq</u>	Saudi Arabia	USSR
5-10%		<u>Pakistan</u>	<u>PRC</u> Nigeria		ROC Mongolia	Portugal	Germany (GDR) USA Gt. Britain
2-4.9%	Burma	<u>India</u> <u>Indonesia</u>	Thailand	South Korea	Brazil Turkey		Germany (FRG) France Sweden
1-1.9%	Afghanistan		Philippines				Switzerland
1%	Bangladesh Nepal		Sri Lanka		Mexico		<u>Japan</u>

* 1974 data derived from Arms Control and Disarmament Agency, _World Military Expenditures and Arms Transfers, 1965-74_ (Washington: U.S. Government Printing Office, 1976), p.6.

infantry forces at relatively low cost; the industrial sector can turn out sufficient weapons for such forces. They can also produce limited quantities of large-scale or advanced weapons—ships, tanks, aircraft, communications, and logistics support facilities—although until recently, this could only be done with external assistance.

It is commonly assumed that because of the extreme poverty of China and India heavy defense spending will be economically dysfunctional, and therefore regarded as deferrable. The limited work concerning Indian military expenditure and economic growth indicates that the opposite may be true. Emile Benoit and his associates determined that there was a positive relationship between defense spending and real economic growth; the most likely explanation being simply that the increased discipline and improved organization associated with the crisis triggering an increase in defense spending also lead to greater efficiencies in the productive civilian economy.[10] In a more efficient or developed economy the relationship may not hold. China's defense burden is much greater than that of India, but there are many ways in which its military units are linked to economic production, and here also the raw figures may not indicate actual defense burdens.[11]

The material base of Japan is quite different from that of China and India but may also qualify it for great power candidacy. Japan's strengths and weaknesses are obvious to even the most casual observer. Already an economic giant, it is nevertheless vulnerable to disruptions in raw materials sources and access to markets for its middle-range technologies. Undoubtedly, Japan could, if it chose to, quickly establish an indigenous arms industry of some sophistication; it has a large well-trained population base.[12]

The Strategic Environment

Unless one is a firm believer in classical geopolitics, there is nothing inherently distinctive or strategic about the location of China, Japan, and India. Their physical existence becomes strategically important only as a function of their relations with the superpowers (especially the USSR), prevailing military technologies, and the time period in which strategic political objectives are pursued. With the advent of nuclear weapons no state can rationally pursue absolute security, but a lesser degree of insecurity can be achieved vis-à-vis threats that are not thermonuclear. What does distinguish these three states is their capacity (or potential capacity) to manage nonthermonuclear military and economic threats from neighbors and both superpowers. We will discuss these threats in greater detail in the chapters on the individual countries as well as below, but first some common features of the three states' strategic environment may be noted.[13]

While they do not share a common external threat or enemy, China, India, and Japan do interact with the superpowers in similar ways, each possessing a particular strategic and military importance. Each country has received military assistance, weapons, or military technology from at least one of the superpowers. Each country has been regarded as a potential (or actual) ally or adversary and has been subjected to pressure from one or more superpowers to join a mutual security pact. Despite this, India and China (unlike other Asian powers) have some ability to resist such pressures and, within the context of alignment, Japan has the ability to retain the option of strategic autonomy. This is a goal utterly beyond the reach of, say, Pakistan, which must stand by while India is courted by the two superpowers.[14] Pakistan is unable to play off one major power against another, for it has lost even its capacity to balance India, retaining only a nuisance value against India. Militarily weak and strategically indefensible, it has been relegated to minor power status for the foreseeable future. Both Koreas, Taiwan, Thailand, and Indonesia are other Asian states with only marginal international bargaining power.

Another shared characteristic of the great powers is the close linkage of status and security. Weak states pursue whatever measure of security that they can achieve within the limits of their resources, and must be prepared to either subordinate their autonomy to powerful neighbors or allies, or engage in an extraordinarily skillful balancing act. For China, India, and Japan, however, the pursuit of status is as important as the pursuit of security. These are states that have a larger conception of purpose than mere survival; their diplomacy is at the service of this broader ideal or ambition. Thus, they may well choose strategies that seem to be self-defeating, needlessly expensive, and that may not even maximize security. One such strategy is the active defense of disputed borders, even at a very high cost. Defending such borders is an act of will or self-assertion; should a state fail to defend itself as in the manner of a great power then it will ultimately lose the status properly accorded to such a power, and its actual security will decline, as neighbors probe for additional weak spots. Smaller or weaker states do not carry this "big power burden."

Finally, these three states share a structural similarity within the hierarchy of nations. Neither global nor regional, they regularly interact with the two superpowers, with smaller neighbors and, in some cases, with each other. These experiences should have sensitized them to the tender feelings of smaller powers and may have done so. (China, of course, has made a fetish of international egalitarianism between large and small states, but the behavior of India and Japan towards their smaller neighbors has also been, on the whole, judicious.) In their different ways each also makes a claim to some form of global influence and each seeks to dominate its immediate region, although in pursuing the latter objective it is vulnerable to the balance of power game. As argued above, one prerequisite of great power status is the capacity to deter outside powers from supporting regional rivals. Much of India's diplomacy revolves around cutting Pakistan's access to U.S., Russian, and Chinese weapons. China seeks to discourage the West and Japan from helping the Soviet Union develop both the advanced and extractive sectors of the USSR's economy. The Japanese have settled for a strategy of dependence upon superpower involvement, but only to keep the other superpower at bay.

Strategic Pressures

China, India, and Japan can all claim security problems that require substantial levels of modern weaponry. Unlike Western Europe, where an alliance structure exists to ensure conventional defense and the involvement of the United States, these states operate in a disorganized and transient strategic environment with little or no regional cooperation. Only Japan can anchor its security policy to a superpower although some circles of American opinion have advocated that Japan pursue a more independent or autonomous policy.

China and India perceive the existence of immediate military threats to their integrity and have armed accordingly. They trust neither superpower and often state so publicly. Furthermore, they are not linked to each other in any coherent strategic framework, and efforts to do so have repeatedly failed.[15] Their policies are likely to be pursued in terms of a strict construction of *national* interest, not regional accord or "stability." Discussions about an independent policy are naturally more muted in Japan, but even supporters of the U.S. tie are aware of and have discussed the alternative of strategic autonomy.

Two areas of military and strategic uncertainty common to all three states are of special interest. One stems from the confused historical pattern of border determination, the other from the logic of regional dominance in a world of superpowers.

China, India, and Japan each have a disputed or threatened border or territory with a power of equal or greater size. These border and territorial disputes, growing out of settlements between earlier regimes as a consequence of nation-creation (or even as the residue of World War II), are of vital importance to the leadership of each state above and beyond its role in the international balance of status and power. They generally have two important internal political consequences. First, India and China have rebellious or incompletely integrated populations on or near their borders. Border disputes are an open invitation to subversion, rebellion, and an assistance to guerrilla movements. Great distances and difficult terrain also facilitate such activities. Second, the leaders of these states are no less vulnerable to the "national security" issue than those of the superpowers, and more so than those of small states with limited military ambitions. Even in Japan, the central leadership must demonstrate continued toughness on territorial questions, which transcend the left-right politics in any nation. Border disputes are a central domestic political issue in India and China.

Security requirements also flow from the intermediate position of these states in the hierarchy of nations. Because each is militarily powerful enough to exert regional influence, their smaller neighbors seek outside assistance in balancing them. This has led to superpower calculations of the wisdom and feasibility of "balancing" China, India, or Japan, versus a strategy of accommodation. In all cases, calculations of small-state survival, middle-state dominance, and superpower competition come into play, leading to a complex multilevel diplomacy. This process is most developed in South Asia, but can be seen in operation in the Gulf region, Northeast Asia, and in superpower diplomacy towards China and its smaller neighbors (especially Taiwan and Vietnam).

Additionally, these three states have reached a level of military power that is of concern to the superpowers themselves, as well as their small neighbors. Because of their concern about regional dominance they have all developed some capacity to resist superpower invervention or raise the costs of that intervention.

This diplomacy of insecurity has produced a ratchet effect in the military inventories of China and India. As they move to meet perceived threats from regional competitors (including each other in some cases), their capacities vis-à-vis smaller regional states become overwhelming, and they enhance their capacity to resist, threaten, or deter the superpowers. The most dramatic example of this is in South Asia. India began an armament program to meet the challenge of Pakistan (which was being supplied with weapons by the United States after 1954). This program helped to some degree in meeting the Chinese threat, and in turn, the major arms acquisition program undertaken after 1962 enabled India to ultimately achieve near-total dominance over Pakistan in 1971. The momentum of that armament program has also given India a substantial capacity to meet naval threats from the sea, a capacity that it did not have during the *Enterprise* mission in the Bay of Bengal.[16]

Image and Reality

> Asie, Asie, Asie,
> Vieux pays merveilleux des contes de nourrice,
> Où dort la fantasie comme une impératrice
> En sa forêt tout emplie de mystère.
>
> —*Asie*, Tristan Klingsor, 1901

Two different conceptual frameworks are of special importance to an understanding of perceptions and misperceptions of China, India, and Japan. The first derives from the false assumption of their "Asian-ness," the second from the role accorded to the larger states of Asia by the classical geopolitical model.[17]

John Steadman, Harold Isaacs, and others have noted that the accuracy of Western images of Asia has not progressed much beyond the romantic poetry of Klingsor or the rough and ready realism of Kipling. Treating Asia as an entity is to engage in myth-creation. As Steadman points out, "The largest significant unit of an Oriental society is not Pan-Asian but 'sub-Asian.'" The term "Asia" designates no single society, as does the term "Europe"; "Eastern," unlike "Western," cannot be employed to identify a single community.[18] Despite this fact, stereotypes of Asia as a whole and of individual Asian states persist at the highest levels of public life.

This is not to deny the numerous ties between India, China, and Japan. Many of these were of great historical importance: Mongol invaders sweeping down through Persia to gain control over the South Asian subcontinent from the thirteenth century onward; earlier, Buddhism being carried from India to China and ultimately to Japan, which was only one element of massive Chinese cultural influence on Japan. These and other connections are of significant interest to the linguist, religious historian, and cultural anthropologist. Added together, however, they are of less *political* importance than the fact of geographic coincidence. Being "of" Asia has often provided an excuse for initiating relationships that were marginal at best; but being "in" Asia has also meant that these states are often viewed by non-Asian powers as having something in common, and they have each had to adjust and readjust to similar stereotypes. Such transient and superficial images of Asian states cover a wide range, but they are notable for their shallowness and their interchangeability. One such convertible pair of stereotypes is the cruel, rapacious, barbaric warrior-horde versus the soft, effeminate, and weak state. Another pair is the industrious, ant-like, mass society versus the indolent, diseased, and lazy peasant. Whole nations are readily pigeonholed and then recategorized. Such stereotypes originated with the contact between expansionist European states and the states of Asia, and grew out of desire to rationalize colonial or hegemonial relationships. The stereotypes made it easier to legitimize the protection of weak Asian states from the strong or (within a region, such as South Asia) the weaker communities from the more warlike and militant.[19] Thus, such stereotypes were applied to relations between as well as within the states of Asia.

A second source of the perceptual errors that punctuate even recent contact between the various nonregional powers and the states of Asia is the dominance of geopolitical models. Whether it is the earlier incarnation as a control over the "heartland," or one of the latter-day versions (such as the zero-sum approach of the cold war), they share one feature: the imposition of a grand strategic pattern on the Asian map, and the reading of that map according to the overlay.

In each case Asian states—unlike their European counterparts—are judged by and large unable either to determine their own interests or to protect them. Their "true" interests are deemed compatible with those of the outside power, an assumption that the Soviet Union's projected Asian collective security system shares with its CENTO and SEATO forerunners. A multipolar model, such as that briefly floated by Henry Kissinger, did recognize the strategic importance and political integrity of two Asian states, China and Japan, and elevated them to the level of Europe, the United States, and the Soviet Union as two of the five great world power centers. But the classic geopolitical model, although sometimes modified to include Japan as a special case, regards the rest of Asia as weak, fragmented, and vulnerable to pressure from the central landmass. It is composed of states with no value except as victims, which require outside protection from the West, whether they wish it or not. Further, the real issue is not their growth, prosperity, or even security but their position in the global tug-of-war. The foreign policies and strategic objectives of these Asian states are thus evaluated within a framework of a superpower struggle. On more than one occasion this superpower involvement has far exceeded any conceivable security or economic value of the state itself, because what is at issue is a broader strategy. As Isaacs notes, the most tragic and prolonged military calamities in Asia have been the result of the interaction of a geopolitical model with stereotypical images of Asia.[20]

The chapters that follow provide ample evidence to show the fallacy of the belief in the incompetence of Asian states in general, and of China, India, and Japan in particular. These are not confused or erratic states—or at least, they are no more confused or erratic than their non-Asian counterparts. Nor are their interests necessarily in harmony with each other or with outside powers, especially in such areas as arms acquisition and nuclear weapons.

It can be argued that an insuperable obstacle to the emergence of China, India, and Japan as great powers is their own domestic weakness. India and China are societies with large ethnic and religious minorities, and they both have enormous regional diversity. India shares with Japan highly pluralist democratic politics. This ethnic, religious, regional, and political diversity could have a restraining influence on foreign policy, at the very least, and may lead to internal dissension, revolt, or even civil war.[21] Thus, some of these states are likely to turn inward, ordering their domestic political affairs before they seek a global or even a regional strategic role.

There is merit to this point of view, and such a course of events cannot be ruled out. Yet despite the growth of regionalism and ethnicity in many states around the world, such a development in India and China still appears unlikely. They seem to have mastered the succession crises that might have provided the opportunity for such a fissioning of society to occur. In India, the most complex of the three states under consideration, the process of accommodation or suppression of subregional and ethnic minorities is particularly well established. Japan, despite its great power and industrial capacity, is the most problematic in terms of the development of a broadly supported, outward looking foreign policy that seeks to match the status acquired as an economic superpower in military and diplomatic arenas. Thus, even acknowledging the obvious domestic problems facing China, India, and Japan, we should not assume that they face inevitable disintegration, weakness, or fragmentation. Nor are their domestic disorders an insuperable barrier to expanded international influence.

Our argument can be summarized at various levels. To begin with, these states each draw from both Western and indigenous models of great power behavior. Their foreign policies are likely to be a synthesis of the two: some degree of emulation of the classical model and some degree of creative adaptation of their own imperial tradition (or cultural memories of that tradition). Furthermore, the three states under examination possess much of the material base associated with powerful states, although the nature of that base varies considerably among them. In each case, however, the critical requirement is that surplus manpower, talent, and funds are generated and can be put to the service of political and strategic objectives. Finally, these states have legitimate security needs, derived from both their location in an unstable Asia and their intermediate position between the superpowers and smaller neighbors. At least two of these states maintain substantial levels of forces in relative terms, and the third, while spending very little on defense on a percentage basis, has great potential for a rapid expansion of arms levels and military technology. Finally, perceptions of these states are subject to more than the usual distortion. Stereotypical images of their capabilities and intentions are widely held and highly transient in content.

Thus India, China, and Japan all share a potential for great power status, and quite possibly near-superpower status. Very few other states in the world fall into this category. Within Europe, one would include Germany and France; outside of Europe, perhaps only Brazil, Nigeria, and possibly Argentina. Each of these states has, or might be able to develop, the capacity to exert dominance in its region. They have or are developing capacities to influence international economic, political, and military systems, and they can each draw upon national traditions and ideologies of imperial grandeur. They represent prime candidates for careful scrutiny as we look forward to the strategic balance of the 1980s.

Toward the 1980s

With respect to the immediate future, there are areas of uncertainty in any judgment about the emergence of China,

India, or Japan as great powers.[22] These include major changes in outside perceptions of these powers, the influence of new military technologies, and the problematic relationship of these three states to arms control and proliferation processes.

Perceptual Discontinuities

A sharper discontinuity between self-image and external (especially superpower) perception involving either China, India, or Japan is likely to have important political consequences. Such a discontinuity is most likely to involve an underestimation of their power, influence, or ambition. This would be one way in which an action-reaction cycle might begin, with the regional power attempting to demonstrate its true interest or capability and the outside power interpreting this as either insignificant (thereby prodding the regional power into further displays of power) or threatening to regional stability (thus challenging the regional power).

There are a number of possible sources of such perceptual errors. One is change in the internal political structure of a state. A vivid example of this was supplied during the nineteen months of Mrs. Gandhi's "emergency." During that period opinion in the West swung away from India in general and Mrs. Gandhi in particular; what was only a feeling of discomfort with India's earlier nuclear explosion turned into active opposition, as Mrs. Gandhi's new regime was now thought to be irresponsible and dangerous.[23] A regime change in Japan would also harden distrust in many sectors in the West, even if (as in the case of Mrs. Gandhi) there were no substantial changes in either nuclear policy or foreign policy.[24]

A second major source of perceptual change would be an exacerbation of international economic problems. Even in the relatively favorable economic climate of recent years, Japan has become increasingly regarded as a state that is inimical to particular Western economic interests. Informed Western observers of Japan, as well, have mixed feelings on this issue. There is not only a feeling that Japan could carry a larger defense burden but also that her small expenditure on defense is an indirect Western subsidy of Japan's remarkable economic growth. There are feelings of envy and admiration for the society that produced that growth, and a sense of being hostage to Japan's implicit threat/capability to rearm. Japanese rearmament might raise strategic problems but it would also slow the process of economic expansion, which is perceived by some as a greater threat than military expansion.

Third, the role of these states in the international strategic system may be further exaggerated or misperceived. Given the desire to reconcile containment with lower defense budgets in the Nixon, Ford, and Carter administrations, it is natural to look to these states as "balancers" of the Soviet Union. To the degree that they perceive their interests and integrity as threatened by the USSR, this is a perfectly sound policy, but traces of impatience have already emerged, especially with Japan and India. In the former case Japan is perceived as a state that fails to carry its full share of the mutual defense burden; in the latter case, India is seen by many as a state that not only fails to "balance" the USSR, but actively consorts with it. The perception of China has been equally vulnerable to rapid change. The present regime or some future one may decide to achieve a formal rapprochement with the USSR; this will in turn be seen as a positive threat to American interests by many, especially if the United States has extended material and military commitments to China.

In all of these perceptions and misperceptions there is a common assumption of the "emerging" or marginal character of the state in question. There is a belief that such states are flawed or in some other way unable to determine their interests. There will be an oscillation in outside perceptions of these states, first exaggerating and then deprecating their might, alternately wishing that they would join the international system as full-fledged powers and fearing that, once part of the system, they would become unmanageable.

Weapons Acquisition and Technological Change

Weapons acquisition can serve many purposes. Aside from legitimate security needs, as defined by a state's own military and political leadership, it is a process that has important economic, political, and symbolic consequences. Of the three states under consideration, one (Japan) can afford to acquire sophisticated weapons in large quantities without evident damage to its own economy. In fact, purchases of foreign weapons by Japan is seen as an important device for easing the balance of payments problem of major Western arms exporters, especially the United States. It is more difficult to judge the economic impact of weapons acquisition by India and China, for their domestic developmental requirements are visible and pressing, although there is some evidence that defense spending is less upsetting to real growth than one would expect.

The political consequences of the weapons acquisition process are no less complicated and equally important. Dependence on foreign weapons sources has proven a costly policy for such diverse states as Pakistan, Iran, Israel, and South Vietnam. Each profited by receipt of weapons from the United States, but each was subjected to pressure from the donor, their patron, or a change in policy that left them with insecure or unreliable support.[25] Yet, as India and China have demonstrated, it is difficult to establish a domestic weapons industry without having it lapse into obsolescence. Japan's strategy is probably optimal: it produces most of its own weapons, but often in association with a foreign manufacturer, thereby gaining the latest in technology while retaining some degree of autonomy. In addition, with Japan's powerful economic infrastructure it can, of course, contemplate autarky more readily. Finally, the symbolic meaning of weapons acquisition should not be ignored. Are India's large forces merely a sop to militant nationalism, symbols of national pride and achievement? Is Japan's light defense burden a product of realpolitik, or a necessary concession to antimilitary and antiwar sentiment in the Japanese population? If, in fact, a weapons program (or the absence of one) is based upon symbolic considerations such as these, it may be no less real than if it grew out of a consideration of security and insecurity, but it certainly might be more transient and changeable.

According to the authors of the individual-country chapters in this volume, the weapons acquisition process is determined largely by rational calculations of security requirements. They tend to support the "political" side in the debate over technological versus political causes of arms races.[26] Weapons themselves are not the cause of conflicts that grow out of clashing interests and ambitions. Nevertheless, the possession of weapons—their numbers, location, disposition, and readiness—can exacerbate such conflicts, and in particular circumstances may indeed be a major contributor to international conflict. The India-Pakistan dispute would be one major case where weapons themselves have assumed a symbolic and political existence beyond their military utility.[27]

An area of particular concern to all three states under examination is the complex effect that new weapons technologies will have on their security and international status.[28] Several new kinds of weapons systems are being introduced into the weapons inventories of the technologically advanced states. Of the three Asian powers we have examined, only Japan is able to even keep up with the research and development in this field.[29] One class of weapon, precision-guided munitions (PGMs), is especially disruptive to conventional military estimates. PGMs are distinguished by their high degree of accuracy (better than .5), their relatively low cost (compared with likely targets such as aircraft, tanks, bridges, and other vulnerable structures), their simplicity of operation, and their versatility. (PGMs can be carried on land, air, and sea-based platforms, although what may realistically distinguish between them is their intended target, not their launch vehicle.) Some hope has been raised that such weapons will tilt the offense-defense balance in favor of the latter, ushering in an era of a military status quo, in which present boundaries and political structures of all sizes are able to defend and protect themselves without threatening their neighbors.[30] It seems more realistic that on land, at least, offensive operations will still be possible—using these defensive weapons—but that they will be more tentative and protracted, with blitzkrieg tactics being discouraged. It also seems likely that massed armed forces are going to be seen less and less frequently as the optimum size for independent operation grows smaller and smaller.

Such small units, combined with very high weapons technologies and a need for advanced communications networks, may in fact work to the overall advantage of the superpowers, at least when facing a state of the second rank. At sea, however, PGMs would work against the fleets of the superpowers. Incursions by major naval units in areas such as the Persian Gulf, the Bay of Bengal, the Sea of Japan, or off the coast of China may become too dangerous when PGMs are in the vicinity. PGMs would seem to have an uncertain effect on the vulnerability of China, India, and Japan to air attack or airspace penetration. Unless a state has access to the latest in PGM technology (which probably means access to one of the superpowers), as well as a very sophisticated air defense command and control network (which India and Japan do have), there would seem to be no effective way of stopping penetration at will by a superpower.

In summary, lacking a consortium of middle-level powers, it would seem to be absolutely vital that such states maintain some ties with either superpower in order to retain access to PGM technology (and to other new technologies, such as laser weapons). Lacking such an association, they will be especially vulnerable to two characteristics of some PGMs noted above: their cheap cost and simplicity of operation. A middlepower not only faces the risk of falling hopelessly behind the superpowers in terms of keeping up (that is, more than one generation of weapons technology in arrears), but also the risk of its smaller neighbors gaining access to such weapons, thus altering regional balances.

Arms Control and Proliferation

It is difficult to estimate with much precision the likely course of the arms control and nuclear proliferation processes in Asia, and the roles of China, India, and Japan with regard to these processes. China, of course, already possesses nuclear weapons, and India has demonstrated a capacity to produce them—halting short of a military program. Japan certainly has the necessary nuclear technology. Several of these states could help others become nuclear weapons powers, or their example might force others to develop such weapons. What is critical is not the capacities of these states to become nuclear weapons powers or to improve the systems that they already possess but the calculations that would support a decision to do so. Before discussing these calculations, which are vitally important yet often misunderstood by outsiders, some terminological clarification is necessary.

Even superficial examination of the problem reveals at least three objectives normally subsumed under the phrase "arms control": prevention, limitation, and control.[31] *Prevention* refers to the initial acquisition of a weapons system, nuclear or conventional. Of the three states in question, China is a nuclear weapons power and India is only a nuclear power, with demonstrated weapons capabilities. *Limitation* as an arms control objective denotes restriction in the qualitative and quantitative levels of weapons. The Indian example of self-restraint after initial acquisition is particularly relevant here. It has demonstrated the ability to build a stockpile of weapons, but has apparently not done so as part of the "nuclear option" strategy.[32] *Control* as an arms control objective refers to the disposition, deployment, use, or command and control arrangements associated with a particular weapons system. Declaratory statements, such as China's "no first use" pronouncements, would fall into this category. Dealing with prevention, limitation, and control can be regarded as separate but necessary components of an arms control arrangement.

Arms control efforts must recognize the diversity of motives behind acquisition. These range from classic arms races (in which nuclear weapons are only a further step in a mutual escalation process) to the impetus of technology (because it can be done, it will be done). In addition, weapons may be acquired for purposes of strategic bargaining with other nuclear powers, or as a symbolic expression of national achievement (directed towards either domestic or foreign audiences). To complicate matters, once acquisition takes place, other considerations may shape levels and control arrangements.

Enough is already known about the politics of nuclear proliferation in these three states to make a few general statements about their role in the proliferation process. First, if India or Japan acquire a military nuclear system, it is likely that the Chinese pattern will be followed. In this case, strategic vulnerability and political conflict with a superpower encouraged the development of a modest retaliatory capacity that was coupled with public assurances of restraint and moderation. None of the three states face a major threat from a smaller neighbor (although their nuclear acquisition might be dysfunctional if it forces states such as South Korea and Pakistan into a nuclear program); it will be their relations with the superpowers that are likely to force them into a military nuclear program.

Obviously, none of these states, including China, can hope to have an effective strategic deterrent against the United States or the Soviet Union for many years without improvements over the kind of system already possessed by the Chinese. A first-strike strategy would be suicidal for them. At the very best they can achieve a limited capacity to destroy a few population centers in retaliation. But the *major* significance of a nuclear capacity for these countries would be the ability to conduct nuclear strikes or demonstrations against the equipment, military facilities, and allies of a nuclear-armed enemy, without being forced to attack major population centers. This form of limited nuclear war is ideally suited as an intermediate strategy for states with great power ambitions; it gives them the status of a nuclear program and permits the development of relevant technologies without unduly provoking the superpowers. Later, missile delivery systems could yield an advanced second-strike capacity, although there will necessarily be a gap between initial acquisition and the establishment of such a secure capacity. This produces a powerful incentive to defer initial acquisition until a rather sophisticated system can be developed—as long as strategic deterrence is handled by some friendly power (the United States in the case of Israel and Japan, and the Soviet Union in the case of India). Maintenance of explicit or tacit guarantees of deterrence must be a high priority for any state serious about acquisition of nuclear weapons in Asia. Withdrawal or weakening of such guarantees is likely to hasten proliferation, although there may come a point when a middle-rank state (probably India) decides that external deterrents are no longer credible or worth the political price, or are even no longer required. In the end, the proliferation of weapons in Asia (and in certain other regions, as well) will depend upon the willingness of the existing nuclear powers to manage their relations with the near-nuclear powers in such a way as to satisfy both the reasonable ambitions and security requirements of the latter. These relations will necessarily include inducements and punishments on behalf of the superpowers, and implicit blackmail—involving the threat to develop nuclear weapons—on the part of the near-nuclear and new nuclear powers. For the latter, the threat may be to expand and refine levels and qualities of weapons or to promote the spread of nuclear weapons to other states and regions.[33]

Conclusion

In his study of the international hierarchy, Steven Spiegel singled out the secondary powers as critical for the future of the international system. Neither superpower is likely to decline, nor are the minor powers and regional states likely to change their role as "ideological and adventurous upstarts." The secondary powers, however, will determine whether or not "an advanced form of the present system continues or whether a considerable transformation occurs."[34] Such a transformation would involve the creation of several additional centers of powers, presumably complete with nuclear forces, exerting influence within one or more regions of the world. A more radical prospect has been advanced by Rajni Kothari, who argues for twenty or more regional centers, smashing the tacit division of the world by the two superpowers.[35] In this plan, states such as India, Japan, and China would naturally dominate their regions, but no superpowers would have worldwide influence.

While we have argued the likely emergence of China, India, and Japan as great or near-great powers, forming a separate and identifiable class of states, this does not necessarily imply concomitant changes in the international system. What may emerge is a "managing agent" world, in which such regionally powerful states as India are conceded local dominance, and superpowers refrain from supporting regional rivals, let alone engaging in direct confrontation. Other states, such as China and Japan, are accorded a specialized global role (political in the case of the former, economic for the latter) without recognition as a full-fledged superpower.

These developments would challenge the superpowers' global strategies, which were based on such general policies as containment, or more recently, the application of ideological, political, or developmental standards. These new middle-rank states would not be seen as prizes in a contest, but as autonomous actors, full components in a restricted geographic or functional sphere of activity.

Awareness of such a change in the international system is bound to come slowly.[36] However, conceding a new status to these states (if, indeed, they seek such change in their international role) is not only compatible with superpower interests, but may be the best way to restrain their military expansion and control the issue of nuclear proliferation. A strategy of preemptive recognition still runs two risks. It may lead to the abandonment of smaller regional states in granting hegemony to powers with regional or subcontinental ambitions. In some cases (such as the Arab states of the Gulf region, or the two Koreas) this may be dangerous and unsettling, in others (Pakistan) it may lead to serious internal disorder. Second, according enhanced status to the emerging great powers may tempt them to reach beyond their capabilities.

As long as the world was dominated by two superpowers, the three major states of Asia have had to function in a relatively hostile world, but one in which there were natural opportunities to exploit the superpowers. The years of bidding and bargaining over the favors of these states may be over, and replaced by a focus on their own capabilities, ambitions, and objectives. Much of our argument implies a certain pessimism about the capacity of the United States and the Soviet Union to accommodate to this new reality. The inevitable influence of distance, perceptual distortion, and ideological obsessions will all make an accurate assessment of these states difficult. It is hoped that the following chapters will to some degree assist in the process of understanding the strengths, weaknesses, ambitions, and limitations of the three major states of Asia, thereby helping global adaptation to their likely emergence over the next decade.

Notes

1. Heinrich von Treitschke, *Politik*, vol. 1, (Leipzig: Max Cornelius, 1897), p. 38. For a careful discussion of the power politics school, see Raymond Aron, *Peace and War* (New York: Praeger, 1968), p. 585 ff.
2. An inventory of these limits is contained in Ellen P. Stern, ed., *The Limits of Military Intervention: Contemporary Dimensions* (Beverly Hills: Sage Publications, 1977). See especially Chapter 7, "Epilogue," by Morris Janowitz.
3. This viewpoint is expressed in a number of books and publications, but most consistently by those of the Stockholm International Peace Research Institute. See, for example, SIPRI, *Armaments and Disarmament in the Nuclear Age: A Handbook* (Stockholm: Almqvist and Wiksell, 1976).
4. Morton A. Kaplan, "Current Issues in European Security" (paper presented to the Chicago Council on Foreign Relations, May 1977), and Ray S. Cline, *World Power Assessment: A Calculus of Strategic Drift* (Washington, D.C.: Center for Strategic and International Studies, 1975). Kaplan writes: "The geopolitical world of which Sir Halford MacKinder once wrote has now become a reality. John Nicholas Spykman's intra-World War II recommendation that only by controlling the peripheral areas could the U.S. contain the central power has become the critical reality of the new international geopolitical system," (p. 1). Also see his contributions in Morton A. Kaplan, ed., *Isolation or Interdependence?* (New York: The Free Press, 1975), especially pp. 15, 31.
5. Cline, *World Power Assessment*, p. 133 ff.
6. There is now a substantial literature on the determination of hierarchy and relative status among nations. For an introduction see Michael D. Wallace, *War and Rank among Nations* (Lexington, Mass.: D.C. Heath, 1973); Steven L. Spiegel, *Dominance and Diversity: The International Hierarchy* (Boston: Little, Brown, 1972); and George Modelski, *World Power Concentrations: Typology, Data, Explanatory Framework* (Morristown, N.J.: General Learning Press, 1974).
7. *Politics and Culture in International History* (Princeton: Princeton University Press, 1960), p. 389.
8. Denis Sinor, quoted in ibid., p. 398. See also Carlo M. Cipolla, *Guns, Sails, and Empires: Technological Innovation and the Early Phases of European Expansion, 1400-1700* (New York: Pantheon, 1965).

9. For the most thorough treatment of the subject see the work of Klaus Knorr, especially *The Power of Nations* (New York: Basic Books, 1975).

10. Emile Benoit, *Defense and Economic Growth in Developing Countries* (Lexington, Mass.: D.C. Heath, 1973), p. 162 ff. See also Henry J. Barbera, *Rich Nations and Poor in Peace and War* (Lexington, Mass.: D.C. Heath, 1973), pp. 121, 126. Barbera concludes that the two total wars of this century "have neither helped nor hindered development or noticeably affected the inequalities between rich nations and poor," although poorer nations may benefit from the integrative effects of war more than richer ones.

11. Military and paramilitary groups take up a wide variety of productive and "nation-building" tasks in China.

12. A useful corrective to the "superstate" syndrome, which has exaggerated this base, is in John K. Emmerson, *Arms, Yen and Power* (Tokyo: Tuttle, 1972).

13. For full-length studies, see Robert A. Scalapino, *Asia and the Road Ahead* (Berkeley: University of California Press, 1975); and Wayne Wilcox, Leo Rose, and Gavin Boyd, eds., *Asia and the International System* (Cambridge, Mass.: Winthrop Publishers, 1972).

14. Pakistan's dilemma is acute and growing worse with each passing year. Several years ago a united Pakistan would have been a candidate for inclusion in this volume as a "near great" state. Today, truncated, it is unable to order its domestic affairs and is at the mercy of India. Recent (June 1977) reports of the withdrawal of French support for a major nuclear program make Pakistan's position even more precarious.

15. For example, the implicit hope in the ACDA project title, "India and Japan: The Emerging Balance of Power in Asia and Opportunities for Arms Control, 1970-75," prepared by Columbia University, 1970. The final project conclusion was that, indeed, there was little if any opportunity for cooperation between India and Japan in terms of the Asian balance of power.

16. The Indian Navy's inventory now includes almost thirty destroyers and frigates, plus a number of patrol boats, many of which are armed with sophisticated ship-to-ship missiles; without its full protective screen such a major ship as the *Enterprise* would run a substantial risk if it were facing a hostile Indian Navy.

17. For an overview of the subject of perception and image see two excellent studies by Robert Jervis, *The Logic of Images in International Relations* (Princeton: Princeton University Press, 1970), and *Perception and Misperception in International Politics*

(Princeton: Princeton University Press, 1976).

18. John Steadman, *The Myth of Asia* (New York: Simon and Schuster, 1969), p. 25; and Harold Isaacs, *Scratches on Our Minds* (New York: Harper, 1958). See also Ignacy Sachs, *The Discovery of the Third World* (Cambridge, Mass.: MIT Press, 1976).

19. This was a favorite theme of many colonial powers even as they recruited their imperial armies from the more warlike and militant groups in several Asian societies. In India, for example, the British often proclaimed their reluctance to depart on the grounds that the Punjab would come to dominate the rest of the subcontinent; this has come to pass in Pakistan, although not in India.

20. "Quarterback Nixon's Asian Game Plan," *The New Republic*, Feb. 19, 1972.

21. This is one of the assumptions of much of the so-called nation building literature. For an excellent study of the politics of predevelopment or underdevelopment versus the politics of development, see Gerald A. Heeger, *The Politics of Underdevelopment* (New York: St. Martin's, 1974); and for the role of foreign policy in such systems see W. Howard Wriggins, *The Ruler's Imperative* (New York: Columbia University Press, 1969), pp. 221-238.

22. We will omit any discussion of such events as crises in political succession, major calamities, floods, famine, and so forth, because these states—even India and China—are capable of managing such crises. However, perceptions of such events by outsiders is important, and will be considered below. Of particular interest is the different ways in which India and China have permitted outsiders to observe the effects of natural calamities.

23. In the words of Rep. Clarence Long, testifying before a Nuclear Regulatory Commission hearing on the supply of enriched uranium to India, "I wouldn't trust any commitment that the Indian government as presently constituted made... India has demonstrated its bad faith . . . India is becoming a police state . . . (it) is practically moving towards a totalitarian police state. For us to be encouraging this sort of development, to be helping it in any way, I think is a reflection on our good faith as a country which claims to be the chief sponsor of democratic regimes in the world." NRC, *Hearing in the Matter of Edlow International Company*, Docket No. 70-2131, Tuesday, July 20, 1976, pp. 21, 22, 26 of transcript. In fact, a decision to ship the uranium was not reached until after Mrs. Gandhi was defeated in 1977, despite the State Department's advice that there was a long-standing commitment to do so (p. 127).

24. There seems to be an emerging consensus on the rationality of

democratic systems which is opposite (but not all that different) from former President Nixon's belief in the relative stabilizing influence of right-wing as opposed to left-wing dictatorships.

25. Pakistan is particularly vulnerable. The United States has effectively ceased supplying major weapons since 1966, although the slack was taken up by China. Now, with an emerging China-India rapprochement and a U.S. refusal to sell even subsonic attack aircraft to Pakistan (the A-7D), Pakistan must attempt to meet its requirements from a European seller (France or Great Britain) or the Soviet Union. The disadvantage of the former is that there are no broader political gains attached to sales, while the disadvantage of the latter is that unacceptable terms (such as agreement to the Indian position in Kashmir) may be imposed. According to G.W. Choudhury, this was in fact the condition for earlier Soviet arms supplies. See *India, Pakistan, Bangladesh, and the Major Powers* (New York: The Free Press, 1975), p. 68.

26. A useful pairing of the two approaches is in Ted Greenwood, Harold A. Feiveson, and Theodore B. Taylor, *Nuclear Proliferation* (New York: McGraw Hill, 1977). See also Ted Greenwood, George W. Rathjens, and Jack Ruina, *Nuclear Power and Weapons Proliferation*, Adelphi Paper no. 130 (London: International Institute for Strategic Studies, 1976).

27. Stephen P. Cohen, "U.S. Weapons and South Asia: A Policy Analysis," *Pacific Affairs* 49 (Spring 1976):49-69.

28. For two surveys see James Digby, *Precision-Guided Weapons*, Adelphi Paper no. 118 (London: International Institute for Strategic Studies, 1975); and Richard Burt, *New Weapons Technologies*, Adelphi Paper no. 126 (London: International Institute for Strategic Studies, 1976). An earlier, broader survey of a future system is contained in several chapters of Nigel Calder, ed., *Unless Peace Comes: A Scientific Forecast of New Weapons* (New York: Viking, 1968).

29. Burt, *New Weapons Technologies*, p. 4.

30. We cannot do justice to the argument in a short space, but George Quester has written an excellent guide to the problem, with a sound historic framework, in *Offense and Defense in the International System* (New York: John Wiley, 1977), p. 183ff. See also Robert Jervis, "Cooperation Under the Security Dilemma" (Center for Arms Control and International Security, UCLA, Working Paper no. 4, April 1977).

31. For a critique of our "arms controller" approach by a fervent advocate of "disarmament," see Alva Myrdal, *The Game of*

Disarmament: How the U.S. and Russia Run the Arms Race (New York: Pantheon, 1976).

32. Laid out in full by Ashok Kapur, *India's Nuclear Option* (New York: Praeger, 1976).

33. Lewis A. Dunn has explored the latter issue in "Nuclear 'Gray Marketeering,'" *International Security* 1 (Winter 1977):107-118.

34. Spiegel, *Dominance and Diversity*, p. 252.

35. See his *Footsteps into the Future* (New York: Free Press, 1975), p. 135ff. and appendix.

36. Michael Wallace's study seems to indicate an order of magnitude of ten to fifteen years between enhancement of capability and status recognition. *War and Rank among Nations*, pp. 52-53.

3
China as a Military Power

Jonathan D. Pollack

Introduction

Few issues have been more pivotal in the history of modern China than those associated with the acquisition and use of military power. For nineteenth-century elites, China's vulnerability to imperialist penetration was viewed principally in terms of the technological proficiency and organizational readiness of Western armed forces, both characteristics sadly lacking in their Chinese counterparts. For twentieth-century leaders, the development of military strength was essential both to China's survival as a political entity and the autonomy of various contenders for power within a fluid and insecure political environment. The twenty-year struggle between Communist and Nationalist forces for the control of China was the logical culmination of these efforts: it was, in purest form, a contest waged through armed strife. The leaders of China's revolution would thus be among the last to deny either the efficacy or necessity of utilizing military means to achieve political ends.

Indeed, the need to maintain and augment armed strength in no way diminished with the onset of Communist rule. While the prescribed ends served by military force changed considerably, the retention of such power was still deemed essential. Mao Tse-tung addressed this issue on the eve of victory in 1949. While clearly relishing the impending triumph of the Chinese Communist Party (CCP), Mao did not consider it an occasion for smugness or relaxation:

> Here, I think it is necessary to call people's attention to the fact that the imperialists and their running dogs, the Chinese reactionaries, will not resign themselves to defeat. They will continue to gang up against the Chinese people in every possible way. . . . They will smuggle their agents into China to sow dissension and make trouble . . . they will incite the Chinese reactionaries, and even throw in their own forces, to blockade China's ports. . . . Furthermore, if they still hanker after adventures, they will send some of their troops to invade and harass China's frontiers; this, too, is not impossible. All this we must take fully into account. Just because we have won victory, we must never relax our vigilance against the frenzied plots for revenge by the imperialists and their running dogs. Whoever relaxes vigilance will disarm himself politically and land himself in a passive position. . . . China must be independent, China must be liberated, China's affairs must be decided and run by the Chinese people themselves, and no further interference, not even the slightest, will be tolerated from any imperialist country.[1]

China's security, Mao further argued some months later, could only be guaranteed by substantially enhancing the nation's military capabilities:

> Our national defense will be consolidated and no imperialists will ever again be allowed to invade our land. Our people's armed forces must be maintained and developed with the heroic and steeled People's Liberation Army as the foundation. We will have not only a powerful army but also a powerful air force and a powerful navy.[2]

Thus, even before the formal establishment of the People's Republic of China (PRC), an unambiguous link had been drawn between the maintenance and improvement of China's armed forces and the protection of the nation's newly won autonomy and territorial integrity. As Mao and other leaders argued, the surest path to reduced vulnerability—if not to absolute security—depended on the continued existence of the instruments of warfare.

Events quickly confirmed these judgments. The outbreak of the Korean war in June 1950 was followed within days by the intercession of vessels from the U.S. Seventh Fleet in the Taiwan Strait, thereby forestalling Chinese plans for an assault on the island of Taiwan, bringing the rapid conclusion of the Chinese civil war. At the same time, the introduction of U.S. ground forces, aircraft, and naval power to the Korean peninsula and the surrounding waters and air space heightened concern in Peking over the security of China's northeastern borders. With the progressive northward movement of U.S. forces, and with Peking's warnings unheeded or ignored by U.S. leaders, Chinese military units intervened massively in the fall of 1950. The Korean conflict was only the first (albeit the most severe) of eleven instances over the next three decades when China's leaders deployed regular military units beyond the territory physically controlled by the People's Republic.[3] Equally significant, the introduction of a hostile and superior military presence near China's borders became a virtual constant in the nation's security environment.

Under such circumstances, measured but substantial efforts to enhance China's defense capabilities have been apparent since the CCP's earliest years in power. A thirty-year Treaty of Friendship, Alliance, and Mutual Assistance with the USSR, signed in February 1950, provided explicit security guarantees to the PRC against external attack. Though now deemed inactive by Chinese decision makers, the treaty did serve as the vehicle for substantial military assistance from the Soviet Union to China. Soviet defense cooperation with China—through equipment transfers, advisory assistance, and ultimately the creation of an entire defense industry—was a central element in the Chinese army's transition to a modern defense force.

While there have been periods when the objective of military modernization has received less emphasis, and in some respects has even been questioned by particular leaders, this goal has now been sustained for nearly three decades. The long-term results of this development effort are readily apparent in the late 1970s. China continues to maintain the world's largest land army (presently numbering more than three million) and an armed militia estimated at seven million.[4] Through a process of unobtrusive if somewhat uneven development and procurement, Peking's existing naval and air forces rank as the world's third largest. The output from China's defense plants also enables the People's Republic to transfer significant quantities of arms to various Third World states.[5] In addition, the acquisition and development of nuclear and thermonuclear weaponry has been a major policy objective for twenty years, with modest but growing delivery capabilities presently in operation. While China's current military strength does not yet come close to rivaling the technological sophistication of U.S. or Soviet power, it does constitute a substantial and significant combat force within East Asia. The growing depth and diversification of these weapons systems clearly testify to China's emergence among the world's major powers.

Since the first years of the new regime, therefore, the enhancement of Chinese military strength has ranked among the principal goals of China's leaders. Yet, the pursuit of this goal has proven difficult, costly, and at times highly divisive. Irrespective of any claims to Chinese uniqueness or particularity, China's leaders have had to confront dilemmas, choices, and constraints comparable to those encountered by political and military planners in other societies. The existence of such conflicts and controversies, and their persistence over time, compels careful reconsideration of China's record as a military power. In particular, approaches describing Chinese security policy as the product of a unitary or monolithic decision process bear closer scrutiny.[6]

This essay is an initial attempt at such reconsideration. In view of the limitations of existing source materials, any such undertaking must be regarded as exploratory. Notwithstanding the importance that Chinese decision makers have long attached to national security, public assessments of this topic have rarely provided the explicitness or detail that definitive analysis would require. For example, it is not possible to discuss in highly refined fashion Chinese conceptions of military doctrine or how such beliefs might vary among relevant policy makers. Publicly available data on the structure of national security decision making is even more unsatisfactory. Nor can the predominant values shaping the thinking of those presently responsible for Chinese defense policy be confidently assessed. Though these informational constraints may now be easing somewhat, the range of available materials remains disconcertingly slim.

Nevertheless, evidence from various sources does enable significant judgments. At least the broad outlines—and in selected instances some of the specifics—of enduring controversies and policy choices can be discerned. Biographic data, for example, permit observations and assessments of institutional affiliations and responsibilities, career patterns, and elite mobility.[7] Military deployments and their variations over time allow conclusions on Chinese perceptions of external threats and the means devised to deal with them. Patterns in the development and production of weapons systems and both the qualitative and quantitative characteristics of such systems provide several significant indications of the competencies and limitations of China's military forces. Finally, and most importantly, there is the cumulative historical record of nearly three decades in power. When judiciously employed, the latter source has produced the richest insights into defense as a political and economic issue.

Three principal concerns—in effect, ongoing organizational dilemmas—have proven pivotal during the course of China's emergence as a modern military power.[8] First is the issue of institutional pluralism. The diversity of institutional roles and functions is symptomatic of the complexity of the military's organizational structure and career hierarchies. The continued vitality of various institutional channels necessarily constrains efforts to devise coherent, broadly applicable organizational norms. On some occasions, it has been the central focus of elite conflict and competition, and could prove so again.

A second key consideration concerns the process of threat assessment and its relationship to military capabilities and doctrine. China's perceived security needs and interests have clearly varied over time. Less settled, however, is the manner and means by which such assessments contribute to the defining of organizational priorities and commitments, both within military organizations and vis-à-vis other political actors. The relationship of extant capabilities and doctrine to broader foreign policy conceptions is yet another possible source of cleavage in this regard.

Finally, policy makers have had to consider the technological and economic basis of defense decision making, both as a recurrent organizational issue and in terms of its relationship to political discourse and debate at other levels. While there have been periods when discussion of this issue has been somewhat muted, this fact should not be deemed evidence that a broad consensus has always prevailed among involved policy elites. Indeed, the significance of this issue may well increase over time.

None of these issues, moreover, appears amenable to easy negotiation or resolution. The effort to devise solutions that are compatible with the spectrum of interests and beliefs discernible on the part of Chinese decision makers constitutes a recurring problem in Chinese military affairs. By considering the interaction of elite values and situational constraints in somewhat greater detail, the dynamics of the Chinese decision process can be more fully illuminated.

The PLA as an Organizational System

The divergent and uneven patterns of military participation in domestic politics over the past decade and a half have substantially revised earlier interpretations of the conduct of the People's Liberation Army (PLA) in the Chinese political process.[9] The variety of scholarly views are far too numerous (and on occasion too obscure) for extended discussion in this context.[10] Consensus, however, does exist for at least one conclusion: military policy cannot be considered the product of a monolithic decision process. As a large, complex organizational system, the PLA has long been encumbered by multiple commitments and responsibilities that overlap, compete, or otherwise conflict. The available data on China's armed forces are insufficient for assembling a comprehensive institutional history, though a recent volume has broken new ground in scholarly understanding.[11] Notwithstanding the gaps and inadequacies in available information, leadership attitudes and practices incontestably vary. The abundance of institutional settings, career channels, and elite perspectives clearly constitute a major problem for coordinating and implementing military policy.

This problem is not a new one for Chinese military elites: multiple forms of military participation in Chinese politics date from the revolution itself. From the earliest years of the CCP's struggle for power, substantial efforts were undertaken to assure that the Red Army performed duties other than fighting. These encompassed tasks such as sideline agricultural production, efforts at mass mobilization, and various administrative and management responsibilities. Through such measures, CCP leaders sought to reduce any possible isolation or estrangement of military units from the populace as a whole. The involvement of both officers and foot soldiers in such tasks was further intended to enhance morale within various units, thereby minimizing possible gaps between the leaders and the ranks. At the same time, involvement in sideline agricultural production reduced the economic burden of the military on society.

While there have been periods since 1949 when the emphasis on such priorities has diminished, these practices have nevertheless remained a major element in the Chinese military ethic.[12] No doubt their durability derives in considerable measure from Mao Tse-tung's strong personal attachment to such policies.[13] Yet, as Mao also recognized, the army's wealth of experience in various nonmilitary tasks (not to mention its recourse to coercive power in extremis) made it ideally suited to undertake various management and internal security functions.[14] As a practical consequence of these considerations, two important conclusions emerge. First, from the earliest months of Communist rule, specific military organizations have continued to devote their energies principally to political, economic, and administrative responsibilities rather than to military affairs per se. Far more importantly, through such involvements particular institutional sectors (e.g., substantial forces at the regional, district, and garrison level, the Production and Construction Corps, the railway and engineering units, and selected elements of the public security apparatus) still maintain and promote organizational norms and career channels largely distinct from the tasks classically associated with national defense.

A cursory review of the distribution of military capabilities in various sectors reveals an additional key fact. The bulk of China's military forces remain infantry units conspicuously lacking in the mobility, firepower, and logistic systems that might render them a fully effective combat force. As such, the virtually exclusive orientation in their military training concerns the most extreme of military contingencies—a conventional invasion by enemy ground forces. As we will consider further in the discussion of threat assessment, such forces are valued principally for their deterrence function. Their capabilities are tangential or even unrelated to the broader spectrum of security needs apparent in various military conflicts involving China since the establishment of the People's Republic.

Yet it is the latter task—the enhancement of military power in relation to potential or actual threats to Chinese security—that constitutes the key recurring theme in the PLA's modernization efforts evident since 1949. Thus, an essential dichotomy between those oriented toward national defense and those concentrating on domestic tasks has long existed within China's armed forces. On the one hand, regular military units and their leaders continue to assume substantial responsibilities in an internal political context. When measured in terms of the distribution of party, governmental, and military leadership positions held by various elites and subelites, the trends over several decades are quite striking. According to one estimate, the number and percentage of positions held by conventional soldiers have risen substantially over time, from slightly more than 26 percent of the available positions in 1956, reaching a peak of 50 percent in 1968 toward the end of the Cultural Revolution, before declining in 1973 to roughly 41 percent.[15]

Yet a parallel trend, if somewhat more uneven over time, has also been evident. An additional group of decision makers, including a number of the PLA's senior leaders, are responsible principally for the external dimension of Chinese military policy. By their known views on military issues and by their institutional affiliations and political conduct, these individuals have generally been much less concerned with the domestic political obligations assumed by many ground force commanders. One is reluctant to describe this group as a de facto national security bureaucracy. Nevertheless, their predominant orientation in military affairs has been to look beyond China's borders. And although these decision makers have hardly been immune to the vicissitudes of Chinese domestic politics, their essential policy commitment to the enhancement of China's defense capabilities has never been in doubt. Indeed, the visibility of these individuals and the media's attention to the issues that have long concerned them have rarely been greater than in the period since the death of Mao Tse-tung and the accession of Hua Kuo-feng to the party chairmanship.[16]

Although membership in such a hypothesized group is hardly fixed, an "inner circle" of approximately a dozen leaders currently appears to exercise principal power and authority in the national security realm. The most prominent individuals include party Vice-Chairmen Yeh Chien-ying and Teng Hsiao-p'ing, the latter serving concurrently as the PLA chief of staff; Minister of National Defense Hsü Hsiang-ch'ien; Marshal Nieh Jung-chen; Politburo member and Commander of the Canton Military Region Hsü Shih-yu; and Vice-Minister of National Defense Su Yü. Infirmities of old age have clearly led some previously influential figures to abdicate their responsibilities (for example, Marshal Liu Po-ch'eng); others, most notably Yeh Chien-ying, are also reducing their activities.

Indeed, the immediately striking fact about these individuals is their age. With the exception of Hsü Shih-yu (born in 1908) and Su Yü (born in 1910), all were born between 1899 and 1903. The death in August 1978 of Lo Jui-ch'ing, former PLA chief of staff and in recent years a member of the Standing Committee of the party's Military Commission, reinforces this conclusion, thinning the ranks of these senior officials even further. Thus, it seems highly unlikely that most members of this group will serve in an active political capacity much beyond the early 1980s. For that reason, less senior leaders have undoubtedly been drawn into this small but vital circle, for example, Deputy Chief of Staff Chang Ai-p'ing. Former military leaders within the state bureaucracy who retain significant responsibilities in the national defense realm, such as Vice-Premier Wang Chen, are also integral participants in this process.

For other leaders much below this upper echelon, the available information on career backgrounds and political perspectives becomes far more scarce. However, the actual identity of younger officials seems less important than recognizing the institutional environments in which they presently serve. If past trends persist, involvement with and responsibility for national security questions will continue to be based heavily on affiliation with several distinct organizational sectors: the General Staff Department, the Ministry of National Defense, the National Defense Science and Technology Commission, the military academies, and those service arms more dependent on high technology weapons.

These judgments, of course, should not be viewed in absolute terms. Official responsibilities in any of the above institutions are no firm guarantee that a particular leader will focus exclusively on issues of national security. Nor should we assume that (to use a hypothetical example) the commander of a given military district or garrison is wholly insulated from developments in advanced weaponry or broader questions of foreign and strategic policy. The available documentation, however, strongly supports the view that the PLA high command has long been sensitive to specialization of leadership functions, and that such institutional arrangements are likely to persist.[17]

Divergent and potentially competing conceptions of military power, therefore, continue to exist within the Chinese armed forces. The internal structure of decision making remains insufficiently understood to consider the possible organizational and political consequences of this dichotomy in much detail. For example, the separation of these tasks may represent a formal division of labor within Chinese military organizations to which all are agreed. Even assuming such a compact, however, it cannot assure unanimity of opinion on resource allocation preferences, assessments of China's defense readiness, or relative receptiveness to external sources of military technology and doctrine. This is amply illustrated by considering one case where conflict between these respective military orientations was particularly acute and visible: the "strategic debate" between Lin Piao and Lo Jui-ch'ing.

During the summer of 1965, Chinese decision makers engaged in a public and unusually explicit debate over the historical circumstances attending the victory over Germany and Japan in World War II.[18] There is still substantial disagreement among scholars on the meaning and significance of this dispute. Most writers believe that Lin (then minister of defense) and Lo (then chief of staff) were embroiled in an indirect debate about possible Chinese involvement in the Indochina War.[19] Lin, it is argued, stressed the degree to which China had depended on its own efforts in defeating Japanese aggression. While outside assistance was always helpful, nations ultimately had to be responsible for their own defense. His analysis could be construed as a veiled but significant signal to leaders in Hanoi that Chinese efforts would remain limited and discrete. Lo, however, placed greater stress on the importance of outside assistance. Thus, he appeared less noncommittal on the issue of possible Chinese intervention, and more inclined to have China assume an active, direct role in the events. Indeed, some of his remarks even appear to suggest that he continued to hope for a joint Sino-Soviet effort to counter the U.S. intervention.[20]

This particular debate was seemingly concluded with the September 3 publication of Lin's essay "Long Live the Victory of People's War."[21] While some U.S. officials took an alarmist view of the document—Dean Rusk, for example, argued that it was "Mao's *Mein Kampf*"—such fears were enormously overstated. Lin's arguments were more appropriately viewed as a suggestion that North Vietnam adopt a policy of military self-reliance. Barring a dramatic shift in the U.S. intervention (for example, an invasion of the North), China was not prepared to fight on Hanoi's behalf. Lo's disappearance from public view in November and his subsequent public humiliation in the earliest phases of the Cultural Revolution further confirmed this conclusion. The case seems an unambiguous instance where two decision makers made widely divergent assessments of a particular external conflict, leading them to advocate very different solutions and postures. The presumed basis of such differences, therefore, rested exclusively on conflicting perceptions and evaluations of an external threat.

A reexamination of both the political context of this controversy and the specific issues under dispute substantially undermines this view. Rather than being an instance of leadership conflict purely in a high-politics vein, Lo's ouster was only indirectly related to inferred differences over Chinese strategy in the Indochina War. Indeed, the argument that Lin was far less inclined toward an interventionist posture was dispelled at the precise time when Lo's views were presumably being rejected. Between September and November of 1965, 35,000 PLA personnel entered North Vietnam for the first time in the war. By the following spring, this number had increased to 50,000. These units, Allen Whiting has noted, "engaged in combat, inflicted losses, and suffered casualties. The antiaircraft divisions fired on American planes and were bombed in return. Engineer and railroad construction batallions, the bulk of the deployment, worked to keep communications routes open despite repeated attack."[22] Thus, there was a limited but credible commitment of Chinese military power to the Vietnam conflict from the earliest days of substantial U.S. escalation.[23] Moreover, Lo's alleged preference for joint defense efforts by Moscow and Peking is utterly belied by authoritative Chinese declarations aired during the early summer of 1965.[24]

Analysis rooted in ongoing issues of organizational and domestic politics yields a far more satisfactory interpretation of these events.[25] Lo's apparent advocacy of a more forceful Chinese response to American escalation occurred at the time of the initial stirrings of the Cultural Revolution. During the early 1960s, Mao had begun to question the political reliability and ideological rectitude of the CCP leadership. He further sensed his growing isolation within the political process. Thus Mao (apparently in need of an institutional base through which to compete with party organizations and in order to reestablish his own political beliefs and preferences) relied increasingly on the PLA in general and Lin Piao in particular.

Lin had had considerable success during these years in resurrecting PLA morale and efficiency, both of which had been badly undermined in the context of the dismissal from office of his predecessor, Marshal P'eng Teh-huai. His success rested on a judicious balancing of professional and political interests within the armed forces. Lin's accomplishments increasingly persuaded Mao to use the army as an emulation model for both party and government institutions and in Chinese society as a whole. Lin and a number of other military leaders, particularly those within the General Political Department, also sanctioned and encouraged increasing PLA involvement in the political and economic area. They further stressed the continued relevance of Mao's writings on guerrilla warfare. By the time of the Cultural Revolution's formal initiation in August 1966, Mao had become closely identified with selected military leaders, and they with him.[26]

Without question, 1965 was a vital year in Chinese domestic politics, and the PLA was centrally involved. Lo's conception of the role and responsibilities of the armed forces, however, ran directly counter to Mao's personal preferences and unmistakable political needs. Progressively during the early 1960s, the chief of staff had assumed a principal role in articulating the interests and beliefs of China's more professionally oriented military commanders. He explicitly advocated increased emphasis on training, accelerated acquisition of advanced military systems, and greater insulation of the armed forces from society. In addition, Lo was subsequently accused of encouraging separatist tendencies within the PLA and even of defying the orders of the CCP Military Commission by allegedly provoking military incidents in the Taiwan Strait.[27]

Under more predictable political circumstances, Lo's actions ran counter to the PLA's ethic of absolute political control. In the context of Mao's severest political test since 1949, his conduct seemed near certain political suicide. The chief of staff still must have calculated that he could marshal sufficient political resources against Lin and the policies he embodied. To the extent that Lo sought to justify his actions by voicing concern over a grave security situation at China's southern border, his tactics made some sense. Quite possibly, however, his advocacy of externally oriented policies at a time when Chinese politics, both military and civilian, were directed inward explains his dismissal from office. In this instance, therefore, divergent beliefs about the most appropriate purposes of Chinese military power directly affected the resolution of leadership conflict. Yet the actual policy outcome—i.e., initiating military deployments to Vietnam that paralleled at least some of Lo's expressed preferences while still ousting him from office—suggests a serious attempt to uphold a balance among competing organizational interests.

These conclusions are further supported by recent commentaries that have dealt (albeit indirectly) with the Lin-Lo conflict. The former chief of staff was among the most prominent military leaders rehabilitated during the mid-1970s. As a major rival of Lin Piao's, his resurrection was no doubt linked to Lin's death and political disgrace. Thus, Lin stands posthumously accused of paying far too little attention to military training—the converse having held for Lo Jui-ch'ing a decade ago.[28] Revisionist accounts of PLA politics during the mid-1960s further assert that Mao Tse-tung and Yeh Chien-ying were directly responsible for launching a mass military training movement in June 1964, but were opposed by Lin Piao, who supposedly complained that "military training has ousted politics."[29] Yet omission reveals as much as commission: neither Vietnam nor the Cultural Revolution is mentioned in any of these articles.

The accuracy of the claims and counterclaims issued over the past dozen years is not a key consideration in this context. It is far more important to recognize the seemingly cyclical alteration of military policies, particularly since the mid-1960s. Such abrupt shifts in the direction and priorities of military planning, though quite possibly overdramatized and overstated in public declarations, suggest a pronounced and enduring duality in institutional structure and doctrine. The existence of such alternative conceptions of military power continues to find concrete expression in the distribution of power within Chinese military organizations and the preferred norms in PLA conduct.

The renewed prominence of externally oriented military policies since the fall of 1976 amply supports this conclusion.

China as a Military Power

Sustained attention to considerations of national defense has undoubtedly been at the expense of a more political conception of military power. This issue has been discussed with unusual candor in a recent *Liberation Army Daily* commentary on the "three ups and three downs" supposedly evident in military training and combat readiness between 1964 and 1975.[30] The implications of these fluctuations were drawn with uncharacteristic bluntness:

> The three ups and downs all center on a fundamental question: what is the army's principal function? The first of our army's three main tasks prescribed by Chairman Mao is to fight. . . . What use is an army if it is not for the purpose of fighting? . . . Had we followed the reactionary fallacies of Lin Piao and the "gang of four," our army . . . would have paid attention to civilian affairs instead of military affairs. . . .
>
> But, how can we "give priority to political education and productive labor and do military training only in the remaining time" and spend most of our time and effort in production and other work at the expense of military training? Those tendencies toward cutting down time assigned to training, making arbitrary assignments merely to mark time and engaging in inefficient labor practices which waste time must be corrected.[31]

Had these alternative policies been followed, the commentary's author observed, China's troops would have become mere "peace soldiers [who] . . . would have acquired no capability to defeat enemies and win victories. The consequences would have led to the gravest dangers imaginable when social-imperialism and imperialism impose war on us."[32]

Thus, Chinese decision makers have long had to consider whether alternative conceptions of the role of military power can ever be adequately reconciled. It has not always proven possible to accommodate the rival demands and expectations that derive from these respective assumptions. International imperatives in particular have tended to pose this issue in stark form. Yet it is not only in the context of threatening external circumstances that such tensions have surfaced. Nor is it simply an unambiguous choice between rival sets of institutional policies. An even more complex range of strategic and policy choices is suggested by the variations in external threats and security requirements apparent since 1949. To consider this issue further, we need to address the overall issue of threat assessment and its relationship to force posture and military planning.

Threat Perception and Chinese Military Power

As noted earlier, Chinese policy makers have confronted a hostile external environment since the earliest months of the People's Republic. But this environment has been far from stable. The identity, severity, and form of external opposition to China have varied enormously, as have Chinese perceptions of and responses to this opposition. Not infrequently, such hostility has been conveyed in military terms. Three questions merit consideration in this regard. When and where have Chinese military units fought, and against whom? What types of military operations have Chinese forces conducted in such instances? And what do overall deployment patterns, weapons systems characteristics, and inferred or explicit conceptions of military doctrine suggest about the forms of warfare against which China is principally prepared?

These questions represent a substantial agenda for investigation. Very little research has focused explicitly on them, largely because Chinese writings have been extremely guarded in discussing such issues. To the extent that authoritative pronouncements bear upon them at all, it is principally through euphemism and indirection. For example, Su Yü, author of one of the few detailed statements on Chinese military policy appearing in recent years, has asserted: "Chairman Mao's guideline on war is, fundamentally speaking, the guideline on people's war.... Future wars against aggression will be people's wars under modern conditions." This latter label could clearly be employed to justify policies and doctrine bearing only superficial resemblance to previous practice:

> [There must be] continuous development of our tactics under new conditions and flexible application of the various methods

of fighting according to objective conditions.... Since war and [its] various stages differ, as do the target, time, area of fighting and the arms and equipment, our method of fighting should change and develop accordingly. We must be flexible in deploying our troops and in using and changing our tactics, and we should constantly study and acquire up-to-date tactics resulting from the development of techniques and equipment.[33]

This passage is far from unique. Other euphemisms abound in official statements: "the requirements for fighting a modern war," "the needs of actual war," and "actual combat requirements" all testify to a recognition of the necessity to adapt to changing circumstances.[34] However, Chinese statements rarely permit definitive judgment on the precise directions and consequences of such organizational needs. Most insights must be drawn from actual Chinese behavior, rather than political or military communications.

Although this review is somewhat exploratory, the principal findings strongly support the opinions expressed in the above quotations. Rather than being confined to a very limited range of possible actions, Chinese military behavior has proven highly adaptive and varied. Very little evidence suggests that China's leaders remain rigidly wedded to a "guerrilla mentality" in the employment of military force. Even a cursory review of Mao's own writings on military affairs reveals his scarcely disguised contempt for mechanistic conceptions of warfare that pay inadequate attention to the conditions and circumstances affecting a specific campaign or battle.[35]

In this regard, PLA commanders for many years have had to assess a substantial variety of military settings, with Chinese forces employing or actively preparing for widely discrepant forms of warfare. Our findings suggest seven overall forms apparent since 1949, sometimes with two or more used in conjunction with each other, and others more restricted in their applicability and usage. Each form, however, requires a distinctive organizational effort in terms of training, military preparedness, and defense capability. Thus, the concrete results apparent in China's military posture and behavior permit more accurate and focused judgment on how PLA planners have decided to allocate their time, energies, and organizational resources.

Main Ground Force Operations

The overwhelming preponderance of China's armed forces consists of infantrymen. Chinese foot soldiers first engaged in significant combat operations during two major campaigns in the Korean conflict, initially in November 1950, and again the following spring.[36] The second principal instance was the border war with India in October 1962.[37] In February 1979, similar operations were undertaken against Vietnam. Literally millions of Chinese soldiers continue to prepare for such contingencies. This continued reliance on a military doctrine deemed hopelessly outmoded by Western observers remains a source of enormous puzzlement outside of China. It is not our intention to try to remove such puzzlement. However, the fact that Chinese forces have engaged in such actions on three separate occasions may be reason enough to explain such preparations.

No matter how implausible such conflicts might appear, they have proven a principal mode of combat operations beyond Chinese soil and in sensitive border areas. Viewed in historical perspective, therefore, the continued maintenance of high readiness in various locales (most notably the Manchurian plain) hardly seems imprudent. Should renewed warfare of a major sort ensue, Chinese strategists will obviously show no reluctance in committing large numbers of troops. In Korea and along the Indian border, for example, Chinese forces were present in division-level strength (thirty divisions numbering 300,000 men in 1950, fifty-seven divisions numbering 570,000 men in 1951, and three divisions totalling 30,000 men in 1962). In the recent Vietnam involvement, upwards of 200,000 troops were committed to Chinese operations. In neither the Korean nor Indian cases, however, did Chinese forces attack in near the strength that such numbers imply. During the Korean campaigns, contrary to assertions of "human wave" tactics, Chinese forces rarely attacked in more than regimental strength, with platoon or company-level actions being far more common. Relying on foot mobility, high troop morale, surprise, deception, and repeated assaults on U.S. outposts and positions, PLA units in the fall of 1950 very nearly succeeded in expelling all U.S. forces from the Korean peninsula. By such tactics, they sought to compensate for the obvious weaknesses of PLA units, i.e., the limits in available weaponry and ammunition, strained logistical lines, and absence of armor and air power to support ground operations.[38]

Indeed, it was only when U.S. forces devised means to thwart Chinese tactics and probe their vulnerabilities that the deficiencies in PLA weaponry, training, and overall organization became fully apparent. By extensively employing air and artillery power against Chinese units, and by initiating counteroffensives against PRC forces, the PLA for the first time in its combat experience was unable to conduct highly mobile warfare. Alexander George has superbly described this transformation and its consequences:

> The arguments employed in troop indoctrination . . . clearly reflected PLA military doctrine . . . that had led to striking successes in the civil war against Nationalist forces. The impression fostered among the troops going into Korea was that they would win as they had won before, with the same weapons and tactics; nothing new was required. . . .
>
> However, by the end of January 1951 . . . the battle line moved forward and backward. Forced to operate in the relatively narrow Korean peninsula the Chinese armies found themselves confronted for the first time by position warfare and a firm battle line. The Eighth Army frustrated all Chinese efforts to create a fluid battlefield situation more suited to exploitation by the guerrilla type tactics employed by the PLA. . . .
>
> By mid-spring 1951, the initial objective of throwing the UN out of Korea by military means was replaced by the hope that a favorable political settlement would be obtained. . . .
>
> It became evident . . . that Mao Tse-tung's doctrine of protracted war could not be successfully applied in the present struggle, which had to be fought on a continuous front in a narrow peninsula against a determined foe who possessed superior weapons and modern equipment. For the first time in their military experience, Chinese Communist leaders found

that lack of naval and air power was a severe handicap to the accomplishment of their military objectives. Under these circumstances ... Chinese combat morale was not able to stand up indefinitely to the strain of a prolonged war outside China's borders that entailed high sacrifices.[39]

If organizations actually learn from their failures and shortcomings, the rout and demoralization of Chinese forces in mid-1951 provide an ideal test case. As always, however, the lessons of warfare are ambiguous. Mao, for one, spoke of the war as a singular victory in which the United States had been defeated.[40] In significant respects, however, the war had a lasting effect on Chinese thinking. First and perhaps foremost, China's Korean experience substantially influenced leadership views on the purposes served by the exercise of military power beyond the nation's borders. Interviews conducted with Chinese prisoners of war unmistakably suggested a pronounced shift in attitude on the part of military commanders. Thus, unlike previous doctrine, Korea was not described as a war of annihilation against one's enemies. Chinese actions were instead deemed necessary but limited measures for compelling the United States to negotiate an acceptable compromise settlement. As Mao himself argued at the time, "We fought U.S. imperialism ... and compelled it to agree to a truce."[41] Public evaluations since the Korean armistice have comparably described Chinese military actions as intended as much for political effect as for tactical advantage.[42]

Suggestive links can also be drawn between Chinese experiences in Korea and subsequent attitudes towards the use of force against militarily superior opponents. While in selected respects Chinese military conduct since the war indicates an aversion to undue risk or provocation, leaders in Peking have not shrunk from using force when circumstances appeared to warrant such action. Yet, if the PLA's resort to force has been fairly frequent, the control evident in such conduct has remained singularly impressive. Particularly when disputes have threatened to involve China in open military conflict with either the United States or the Soviet Union, Chinese actions have been carefully limited in scope, intensity, and duration. Moreover, increases in troop deployments to particular areas where tension has risen have often been preceded by high level efforts at deterring or defusing incipient conflicts. In addition, force has been employed almost exclusively in the context of asserting claims to territory disputed by China and other states. Chinese military conduct cannot be deemed part of a larger design of expansionism or encroachment, although the recent episode with Vietnam suggests an increased willingness to employ force for political effect beyond China's territorial boundaries. Nevertheless, Mao's private assertion during 1969 has been generally applicable to Chinese conduct: "If the enemy should invade our country, we would refrain from invading his country. As a general rule, we do not fight outside of our country."[43]

It is doubtful, however, that all leaders derived comparable lessons from such a singular experience. Certain institutional sectors were seemingly more affected by Korea than others, in particular those associated with the PLA Air Force. Indeed, notwithstanding the staggering losses encountered by Chinese forces in the spring of 1951, the overall agenda for infantry modernization that a ground force officer could have plausibly developed at that time remains substantially unfinished more than a quarter century later. PLA ground forces remain woefully deficient (in comparison to their potential adversaries) in overall firepower, in the extent of mechanization and other factors central to effective infantry operations and are subject to the vulnerability of Chinese logistic systems.[44] Yet such assessments pay insufficient heed to the form and locale of warfare that Chinese military commanders expect to encounter. Chinese forces have never again found themselves engaged in combat operations of such magnitude, duration, and severity. Nor has any opposing military force undertaken significant operations in areas over which the PRC held unquestioned territorial sovereignty. Thus, the value of infantry forces purely at the level of deterrence must be explicitly recognized. (This issue will be considered further below.)

This is not to suggest that Chinese forces experienced no significant transformation after their Korean debacle. Indeed, as the infantry operations against India amply demonstrate, Chinese units evidenced impressive capabilities in all phases of the 1962 campaign. While the scale of the latter conflict and the capabilities of Indian forces were hardly comparable to those encountered by the PLA in Korea, the 1962 successes nevertheless suggest planning, preparations, and readiness ideally suited to the conflict at hand. Chinese forces again undertook only small unit operations, never numbering more than several thousand. Preassault use of artillery power was devastatingly accurate and effective. Notwithstanding the extremely attenuated logistical lines and the inhospitable terrain, Chinese forces never lacked for any essential equipment or supplies. Actual operations—taken in full view of beleaguered and ill-equipped Indian units—were initiated with enormous superiority in favor of Chinese forces. Yet Chinese military actions remained highly restrained, being closely coordinated with diplomatic activity at every phase of the operation, even to the point of the PLA withdrawing from all captured territories. It was, in effect, the quintessential limited war operation undertaken in close conjunction with vital but constrained foreign policy objectives. No doubt the benefits inherent in the total military control achieved by Chinese units in 1962 have not been lost upon either Chinese or Indian elites.

Small Unit Border Operations

This level of military activity is clearly much less severe, yet it constitutes a key element in Chinese defense planning. There is a long history of such units being deployed to various sensitive, geographically remote border areas where territorial jurisdiction remains contested. The presence of these forces is frequently a routine operation linked to lower levels of tension and threat perception. Yet these units (usually lightly armed infantrymen) have found themselves engaged in combat in periodic instances along both the Sino-Indian and Sino-Soviet borders. Unlike the full-scale deployment of regular infantry units, however, these forces as a general rule have not had recourse to significant logistic support or back-up personnel. Their presence is deemed necessary for purposes of reconnaissance and for asserting Chinese territorial claims at times when tensions may be rising. When and if conflicts have ensued with rival border patrols, it is principally their own battle to fight. Thus, the escalating tensions along the Indian border in 1962 were stimulated by the presence of border units on both sides. But it is also clear that Chinese forces had increased both their number and activities in a highly visible manner, thereby conveying (in a way that press commentary or diplomatic protest could not) the seriousness with which the border situation was viewed. It was only in subsequent weeks (as Chinese deterrence efforts went unheeded) that main ground force personnel were introduced, virtually assuring the onset of full-scale hostilities.[45]

The Sino-Soviet border clashes of 1969 offer a parallel to the Indian case.[46] The initial armed hostilities on March 2 had been preceded by confrontations and incidents dating from late 1967. By the time of the exchanges of early March, Chinese forces (ignoring Soviet warnings) had doubled the number of troops undertaking patrols in and around Chenpao (Damansky) Island. As in 1962, PLA units relied on superior concentration of forces, advantageous locations for observations of Soviet units and unhampered movement of Chinese personnel, and some measure of surprise in order to dislodge or counter the presence of opposing forces. Such conditions, however, no longer existed in the second principal clash (March 15): Soviet forces had vastly reinforced their troop strength, firepower, and mobility, which contributed to far higher casualties and less conclusive results than in the first exchanges. Unlike the Indian case, however, a broader conflict did not ensue.

It is clear, therefore, that actions of border forces, in particular at times of growing tension, are best viewed as an adjunct to diplomatic and political activity at other levels. Chinese actions along the Indian and Soviet fronts were both related to the control of isolated pieces of territory. In no sense could they be construed as intended principally to annihilate (or compel the surrender of) all opposing forces. As a signal of Chinese determination to demonstrate the seriousness of particular territorial claims and to confront perceived external threats, they were indeed effective. Yet, as both principal instances suggest, they may not have produced the preferred results, but instead contributed to raising the political and military stakes for both sides.[47] Thus, while proving effective and wholly necessary in a tactical sense, they may well have increased the probability of far larger conflicts.

Coastal Defense Operations

Not all threats to Chinese territorial sovereignty have originated in remote border areas. Along China's southeastern coast, substantial efforts have been made since the PRC's earliest years to counter espionage, sabotage, and harassment operations undertaken by Chinese Nationalist forces. In coordination with militia units and navy patrol craft, PLA troops in coastal provinces such as Fukien and Chekiang have sought to deny Taipei's forces the opportunity to subvert local authority, harass shipping and other commercial activities, or otherwise interfere with efforts at economic development in the coastal areas. The conception of joint defense efforts by various military sectors—including the involvement of local paramilitary units—has been followed in both theory and practice.[48] The mission is an ongoing one involving the coordination of considerable numbers of local personnel as well as provincial ground force units. At times of greater tension and domestic vulnerability (for example, the spring and summer of 1962), military authorities have not hesitated to increase substantially the numbers and activities of PLA personnel involved in such operations. Thus, in the first three weeks of June 1962, more than 100,000 troops were transferred to Fukien to deter what was perceived as a possible Nationalist invasion effort against the coastal provinces.[49]

The task of coastal defense, then, is surely an important one. It enables substantial usage of provincial-level army and navy personnel with less access to advanced weaponry and greater responsiveness to involvement in civilian responsibilities, thereby enabling main force units to concentrate their energies and efforts at other levels. It also lends substance to the constantly reiterated need for close interdependence between armed forces and society without interfering with larger national defense operations. The reliance on the militia, however, is in no way a guerrilla strategy of "luring deep": rather it reflects the plausible and necessary mission of protecting China's coastal areas from intrusion and interference.

Coordinated Air and Sea Operations

Yet another form of warfare is also discernible along China's coastal provinces and territorial waters, and it is one that cannot be deemed a defensive mode of operation. The continued and (in selected cases) substantial presence of Chinese Nationalist forces on various offshore islands during the 1950s gave both air and naval forces a major responsibility and opportunity to undertake offensive military operations. These operations involved the coordinated use of air and naval capabilities, sometimes also including amphibious operations and heavy reliance on artillery. Major operations involving one or more of these forms of combat were undertaken against Hainan and Quemoy in 1949 and 1950, against Yikiangshan and the Tachens in 1954 and 1955, and again against Quemoy in 1958.[50] Through overwhelming concentration of forces and surprise in timing and execution, efforts at quick, decisive action either to rout opposing forces or compel their surrender could be undertaken. The locale and military requirements of such operations clearly have no precedent in the PLA's pre-1949 history. Thus, the unevenness in the PLA's overall record in such operations is hardly surprising, with conspicuous successes as well as disastrous failures. But there is little denying that such tasks constituted a major preoccupation of military leadership and personnel during much of the 1950s, necessitating the creation and maintenance of an entire air and naval base structure and related facilities in coastal areas.

Nor has this style of operations passed from view with the apparent quiescence of the offshore question since the early 1960s. Quite the contrary: heightened involvement in precisely such activities is again very much apparent. Thus, Chinese decision makers have in recent years evidenced increasing concern about the control of various contested island chains in the South China Sea, some hundreds of miles distant from the mainland.[51] Asserting claims to such islands requires the mobilization of organizational effort demonstrably at variance with PLA traditions. In no other area is the need to progressively develop an externally oriented conception of military power more apparent. Indeed, to press such territorial claims to their fullest extent—not to mention to assert complete control of a 200-mile economic zone—implies a long-term development program oriented toward a "blue water" navy.

Although present trends are far from definitive, several indications point strongly in the direction of a major emphasis on naval capabilities. Chinese naval power, no matter how outmoded some deem it in comparison to American and Soviet naval strength, experienced substantial growth during the latter half of the 1960s and first half of the 1970s. The discriminating development of capabilities geared to PRC needs—for example, a fleet of more than 60 diesel-powered submarines, growing numbers of missile boats, and a few destroyers—has substantially raised the costs for any enemy encroachments on Chinese territorial waters.[52]

At the same time, Chinese training programs and actual military operations suggest a progressive, long-term effort at enhancing the PRC's ability to project naval power beyond Chinese shores. The decisive air and sea operations against outmanned South Vietnamese forces on the Paracel Islands in January 1974 offered the first visible evidence of such capacities.[53] Additional reports of joint air-sea-amphibious maneuvers along China's southeastern coast during July 1976 and August 1977 lend further credence to these judgments.[54] Indeed, one cannot indefinitely dismiss the larger issue of Taiwan from such calculations.[55]

No matter how credible various territorial claims might appear, however, the military requirements for these situations can hardly be deemed defensive. This fact is hardly altered by articles justifying naval development in the context of defensive tasks and international imperatives.[56] The basis for a substantially different range of organizational policies and practices has thus been created, and could well see heightened attention in future years.

Linear Defense: The Protection of Airspace and Territory

Not all the possible uses of China's modern military power should be deemed offensive either in intent or execution. From the time of the PLA's initial exposure to the uses of airpower in the Korean War, there has been considerable awareness of (if uneven attention to) the potential vulnerability of Chinese airspace to enemy penetration. Measures to counter such vulnerabilities have been apparent ever since, particularly at times when military actions in Chinese airspace have been a distinct possibility. These measures run undeniably contrary to assertions of a people's war strategy of "luring the enemy in deep." In seeking to protect Chinese territory against possible intrusions, the necessity of upholding sovereignty over China's land, air, and sea has been identified as a valid and necessary national security objective. What do the available means and measures employed in defense of such a principle suggest about the seriousness with which this goal has been pursued?

There have been three major instances when the protection of Chinese airspace has been of particular concern to China's leaders: Korea in the early 1950s, the Fukien Front in 1958, and the Sino-Vietnamese border in the mid-1960s. In each of these cases, however, the intent to uphold control of airspace has conflicted with other, equally pressing concerns. Specifically, one has sought to assure that Chinese defensive operations not be judged unduly provocative or risk-taking. Thus, rigorous limits have been upheld in the scope of air operations so as to avoid the onset of any wider conflict. Independent of this constraint, however, the overall record of Chinese air defense units does not seem particularly impressive or encouraging. In Korea, a modern air force was virtually assembled overnight; many of those flying aircraft with Chinese markings were reportedly Soviet pilots.[57] Under such circumstances and without previous exposure to aerial combat, PLA air units sustained enormous losses.[58]

The events in and around Quemoy in 1958 seem a more appropriate and reasonable test of Chinese capabilities. Flying aircraft comparable or virtually equivalent to Chinese Nationalist jets, PLA jets fared very poorly, sustaining major losses in a number of embarrassing aerial exchanges. Even acknowledging that Chinese forces were under rigorous constraints not to expand the scope of the conflict (thereby risking possible engagements with U.S. aircraft in the area), their overall combat performance was not likely to instill confidence in military commanders that their airspace was inviolate.[59] Chinese behavior in and around the Indochina border was also extremely circumspect. Although antiaircraft units did not shrink from firing at U.S. aircraft, Chinese leaders avoided any undue reactions to the occasional overflights of PRC territory. As in 1958, governmental agencies limited themselves largely to verbal protests about U.S. military activity that ultimately proved so ritualized as to be virtually devoid of operational meaning.

A rather different and more varied challenge to the protection of Chinese territory has been posed by the Soviet military presence near China's northeastern borders.[60] The multifaceted nature of this presence seemingly confronts Chinese commanders with impossibly difficult defense tasks should Soviet forces invade Chinese soil. PLA deficiencies in firepower, mobility, antitank weaponry, tactical aircraft, and the like appear to leave the North China plain extremely vulnerable to a combined Soviet ground-air assault. Within the limits of Chinese equipment and personnel, however, PLA commanders have sought to counter such vulnerabilities. Approximately two-thirds of China's front-line infantry divisons are located in the Peking and Shenyang Military Regions.[61] The bulk of China's tanks and armored personnel carriers is also found in these regions. Similarly, as tensions have diminished along China's southeastern coast, Chinese tactical aircraft (and some measure of troop strength) have been redeployed to the North.[62] Significant efforts to upgrade antiaircraft operations and the like are also underway. We cannot conclude, therefore, that Chinese forces are prepared to yield their northernmost territories without a significant frontal battle.

While increasing emphasis is being devoted to these precise defense tasks, major doubts remain about their adequacy in the face of the enormous superiority of Soviet forces. The predominant view in the West is of extreme Chinese vulnerability.[63] In certain respects, such concerns are increasingly shared by Chinese, as well.[64] Notwithstanding these concerns, however, Chinese commanders still insist that Soviet forces will ultimately have to fight the type of war that their Chinese counterparts want to see fought.[65] Yet such expressions of confidence are premised on a belief that the final form of a Soviet attack must be an all-out effort to subjugate and occupy the PRC. China's efforts in linear defense, however, suggest if not contrary judgments, then at least an alternative policy and military posture.

Can this disparity be reconciled? Chinese statements of ultimate confidence in the high morale and inexhaustible reserves of infantrymen in hand-to-hand combat represent the irreducible core logic in Chinese military strategy: no one rational will ever again commit an army to invade China. The fact that war has not occurred during a full decade of acute military tension (and extreme Chinese vulnerability) further confirms this belief. It also suggests that selected dimensions of Chinese military power (and the doctrines associated with them) are intended at least as much to deter armed conflict as to repulse China's enemies. This becomes even clearer when two final levels of organizational effort are considered: nuclear deterrence and people's war.

China's Nuclear Deterrent

No Chinese military program has involved more expenditure and effort than the development and deployment of nuclear and thermonuclear weaponry. The nuclear issue is far too complex to be discussed at any length in this essay.[66] We will instead briefly summarize what the premises, expectations, and accomplishments of the program to date reveal about Chinese conceptions of the uses and consequences of nuclear weaponry.

To the extent that Chinese decision makers have discussed such capabilities publicly, they have attached virtually no military utility to their acquisition. Nuclear weapons have generally been deemed suitable only for purposes of intimidation, and for countering any attempts at blackmail. Whenever their possible military use has been discussed, Chinese spokesmen focus in particular on their limited relevance to any hypothetical conflict involving China, given the country's continued effort at dispersal of population and industry. As one recent article has asserted: "Everybody knows that under the conditions when both sides have nuclear weapons, such weapons pose a much greater threat to the imperialist and social-imperialist countries whose industries and population are highly concentrated."[67] Any effort to assert the decisiveness of nuclear weapons has been comparably challenged.[68] Yet none of these analyses suggest that nuclear weapons would not be employed against China; several (including the one cited above) specifically note that China could and would respond by employing its own nuclear weaponry against anyone launching such an attack. Given such views, China's nuclear program retains powerful advocates in Peking, with calls regularly heard for an expanded and more sophisticated strategic arsenal.

Trends in PRC nuclear deployment since the late 1960s, however, still suggest a program restrained in size and intended almost exclusively to deal with only the most extreme of contingencies—i.e., as a limited means of retaliation against the prospect of any irrational nuclear attack on China. Such limits suggest two possible considerations. First, the nuclear program and the technological and economic requirements associated with it place it in direct conflict not only with civilian economic goals, but with other aspects of China's military modernization effort.[69] Second, Chinese decision makers remain acutely aware of the possible consequences of a larger nuclear program for China's relations with other states in East and South Asia. Thus, there is no evidence that Chinese leaders have sought to parlay their strategic forces for political advantage vis-à-vis China's nonnuclear neighbors, such as Japan. While the program has clearly served broad national objectives, and was so construed by Mao personally,[70] it has never been suggested that the nuclear program provides any leverage or advantage either for Chinese diplomacy or for PLA commanders.

China's nuclear deterrent, however, continues to rest largely upon delivery systems of Soviet design, some more than two decades old. The possible vulnerability of these systems—e.g., intermediate-range manned bombers and liquid-fueled, soft-site missiles—remains a key issue. Chinese defense planners have sought to mitigate this problem by efforts at dispersal, camouflage, and mobility. Notwithstanding such efforts, particular organizational constituencies remain unpersuaded that the quality and quantity of present Chinese systems are sufficient, even (and especially so) for purposes of deterrence. And, beyond their utility at a strategic level, such capabilities also could bear upon battlefield conflicts. For example, some of China's shorter-range missiles are deployed so far into the interior that (even when fired to maximum range) they could only be used against targets on Chinese territory.[71]

Thus, there are substantial institutional pressures for China to mount a nuclear effort far more varied, intensive, and expensive than has yet been undertaken. Up to the present, however, the dominant conception has remained one of pure or finite deterrence. Specifically, Chinese officials have no desire to employ such weapons, but nevertheless require them to cope with the most extreme (if highly unlikely) of national emergencies. In view of the broad range of defense needs for Chinese forces, it seems highly probable that such a conception will continue to prevail. Even a greatly expanded effort could still be construed exclusively within the context of a deterrent rather than war-fighting posture.

People's War

At the other end of the technological and economic spectrum of Chinese deterrence efforts is the doctrine of people's war. Notwithstanding the considerable attention and controversy surrounding this concept, the reasons for its reemergence as a political and military strategy during the mid-1960s remain dimly understood. No matter how grandiose the conceptions that have occasionally been presented in defense of people's war,[72] several conclusions remain incontestable. First, in virtually every instance since 1949 where the PLA has deployed or used force, China's armed forces have relied on means of warfare decidedly different from any putative "guerrilla model." On those occasions where a substantial emphasis has been devoted either to mass militia strategies or to the concept of people's war more explicitly, it has been as an adjunct to, rather than the principal component of, Chinese defense strategy.

Indeed, reliance on a people's war strategy does not genuinely constitute "worst case" thinking. On the contrary, for those decision makers identified with people's war as an organizational doctrine, it is "best case" thinking—representing a mode of warfare that no reasonable or rational adversary could possibly contemplate. It is, in effect, the war one will never have to fight. To the extent that it has been emphasized at the expense of other more limited (if more probable) forms of warfare, it is quite possibly reflective of the duality in Chinese military thinking discussed elsewhere in this essay.

Yet, as a means of instilling confidence at times when pessimism about external threats has been particularly acute, its efficacy has been undeniable. As an editorial in *Chieh-fang-chün Pao* has recently noted:

> The Chinese people must defend their territory and sovereignty against infringement. . . . [Thus] not only must we have a powerful regular army, we must also organize contingents of the people's militia on a large scale. If everyone of the 800 million people knows how to shoot and fight, forming a gigantic net over our vast land—this in itself is a powerful protection of China's security and a stern warning to social-imperialist and imperialist aggressors.[73]

With due allowance for overstatement, the above citation conveys how such an allegedly outmoded conception of military strategy can still be held to serve the security of China.

People's war, then, constitutes only one component of a varied and complex set of organizational policies. Nothing in our discussion of defense strategy suggests that China's military planners subscribe to any mystical, dogmatic belief in past organizational glories. The entire record of China as a military power points to a very contrary conclusion.

It is clear, therefore, that Chinese behavior reveals a considerable variety of organizational responses to external threats. To be sure, these responses are frequently obscured by official expressions of policy that rarely if ever allude to such variations. When and how Chinese policy makers have tried to reconcile (or at least balance) these divergent institutional practices and policies are beyond the scope of this essay. However, one level of consideration has undeniably been a recurrent problem, and not only for those responsible for national security per se. How are such compelling national needs to be met within the context of a largely agrarian and only partially modernized economy and society?

Technology, Industry, and National Power

Decisions related to national defense, for China as for all states, are ultimately ones of allocation. Toward what ends, and with what degree of effort, are organizational capabilities most appropriately devoted? What amount of expenditure does a given security need merit or require, and how should alternative choices be weighed? If a certain defense need seems particularly acute, what consequences does this have for competing budgetary requirements, whether military or nonmilitary?

While each of these questions are central to any allocation decision, the actual process of determining "how much is enough" is not nearly as rationalized or coolly analytical as the label seems to imply. Especially for an economy at China's overall level of technological and industrial development, the more appropriate focus is on constraints, not choices. All too often, scholars ignore the obvious limits of Chinese power that cannot be overcome without prodigious, long-term effort. The sheer size of China, and the significance of its capabilities in absolute terms, obscure a central fact: China remains a labor-intensive economy with a growing but relatively small advanced industrial sector. Its power must still be viewed more in terms of potentialities than achievements. Such considerations are directly reflected in China's present defense capacities. A U.S. governmental report has effectively described this relationship: "The Chinese military in many ways mirrors the economy that supports it. For most of its combat strength, the greater part of China's armed forces relies upon manpower and easily manufactured, low-technology weaponry. This like most of China's economy is labor-intensive, with little capital."[74]

Notwithstanding such limitations, China's achievements in the national defense realm remain considerable. According to U.S. data, Chinese military expenditures in 1976 (in 1975 constant dollars) totaled $32.8 billion, the third highest figure globally.[75] No doubt this figure in part reflects the sheer size of China's armed forces, further indicating what even incremental growth accounts for, given a military force of such enormous absolute numbers. But it also suggests how serious a preoccupation national defense has been for China and (given the renewed attention to military modernization in recent years) how likely it seems for national defense to remain so in the indefinite future.

China's armed forces, however, will only progress as far as available industrial, technological, and budgetary capacities permit. This in particular holds true for defense sectors that are judged especially deficient or outmoded—i.e., those that depend on advanced technology where China's economy is weakest. Indeed, U.S. government sources believe that the proportion of China's advanced industrial sector committed to defense production already is "far larger . . . than is the case in the U.S. or USSR."[76]

Such considerations lend some perspective to the obvious dilemmas and difficulties that face China in terms of basic investment strategies and specific allocation priorities. No matter how essential particular defense needs might be judged, they can exert a deleterious effect on China's overall effort at modernization and industrialization. At the same time, efforts to simply purchase completed weaponry from abroad have generally been perceived as unwise for two reasons: the potential costs involved in terms of available currency reserves, and the possible complications and dependency such purchases might well engender both politically and technologically. In an overall sense, there has been a definite preference for indigenous manufacture of defense items. Yet, in so doing the items one is able to produce will necessarily reflect the overall level of China's technological competence, which obviously lags well behind China's principal political and military rivals. How have policy makers attempted to minimize the potentially severe consequences of a large military budget without unduly jeopardizing Chinese security?

A full accounting of this issue is well beyond the scope of this essay. China's leaders (both military and civilian) have not been wholly forthcoming about such a sensitive question. Nevertheless, it is possible to reconstruct, at least in a partial sense, the solutions that policy makers have attempted to devise. First, Chinese planners have obviously recognized that there is a logical progression in the indigenous development of competencies required to create autonomous defense industries and programs. Attention must be devoted to a number of realms. The reason for this is more than simply to gain access to specific equipment and defense items.[77] For a full-scale industry to develop, one must acquire (and gain practical experience with) a broad range of technological, engineering, and manufacturing skills. A scientific and management infrastructure for research, development, and production must be assembled. To fully equip and maintain a modern military force will also require the training of military personnel to use such equipment in an appropriate manner, and the necessary technical expertise to maintain, repair, or otherwise refurbish modern weaponry.

Thus, the overall level and degree of competence achieved in these realms are a good measure of the extent of a given nation's independence in national defense production, not to mention in industrial development more generally. By such criteria, China has advanced further toward military self-sufficiency than any other Third World state. Though it is fashionable to dismiss Chinese weapons as antiquated and militarily suspect, such commentaries lose sight of the PRC's singular achievements. First, in selected realms and for particular needs, Chinese factories produce more than representative military equipment. The AK-47 rifle, for example, is widely considered among the finest infantry weapons in the world. Second, regardless of the shortcomings of specific weapons, it is hardly insubstantial (either militarily or politically) that China currently possesses an entirely autonomous network of defense industries across a broad spectrum of military needs.

The significance of this latter point becomes clearer by a

quick review of China's progress in these areas since the establishment of the People's Republic. In 1949, Chinese weapons inventories consisted exclusively of captured Japanese, American, and Kuomintang stocks, combined with whatever production domestic factories could furnish and modest amounts of Soviet aid. Through grants, transfers, and purchases during the Korean war, China rapidly began the transformation to a modern national defense force—with especially pronounced results in the creation of an air force and modern infantry and armored units. Following the end of the Korean war, attention turned to the creation of manufacturing facilities under Soviet license. By the late 1950s, these plants were producing military equipment in virtually all categories of need, including jet aircraft. With the abrupt cessation and withdrawal of Soviet advisory assistance in 1960, Chinese scientists and engineers had to quickly undertake independent management of all arms plants. Though serious setbacks and difficulties were encountered, China had by the mid- to late-1960s resumed production in all key defense facilities—and obviously independent of foreign management or control. Thus, by this time China had achieved self-sufficiency (if not initiative) in most areas of defense manufacture. This included the ability to "back engineer" key weapons systems for which only prototypes or limited supplies were available, such as the TU-16 intermediate bomber.[78]

To move beyond such production capacities and toward a self-sustaining design and manufacturing effort is far more difficult, and only modest beginnings have been made in this realm. The first such step is the ability to undertake modifications and improvement of preexisting designs. This goal has been pursued by Chinese scientists and engineers in certain areas of defense production since the late 1960s, but with very uneven results. The felt urgency to break free of past restraints and demonstrate an ability to undertake autonomous design and manufacture is understandable in the context of asserting national independence, but not easily realized. China's experience with the F-9, the nation's first domestically designed and produced fighter-aircraft, offers an instructive example. Since first appearing in 1970, the aircraft has been produced in only limited numbers, and is judged a failure or only partial success by many military observers. Indeed, whether the aircraft should be deemed wholly Chinese or simply a modified MiG-19 remains an open question.[79] Such problems illustrate the long-term effects of technological dependence, given the disparate, highly complex skills that are called upon in the manufacture of sophisticated weaponry.

If there are great difficulties in moving beyond existing models and systems, the ability to truly engage in an indigenous design and production effort becomes clearer. This level of scientific and technological competence still remains the exclusive preserve of the major industrial powers, and seems certain to remain so for the foreseeable future. Chinese military planners are under no illusions about the potential sources of technology for their current effort to upgrade the nation's defense capabilities. Now, as in the past, China must look abroad.

But on what basis—political, managerial, or economic—are such technology transfers to be undertaken? For Chinese decision makers, the maintenance of indigenous control remains a paramount consideration. Given the enormity of China's needs and the sheer size of its armed forces, the outright purchase of weapons from abroad, even on an extended credit basis, makes little sense as a long-term policy. In addition, undue reliance on grants, purchases, and transfers still leaves one potentially vulnerable to the vagaries of the supplier state's policies and capacities. Rather than risk such dependence, one must attempt whenever possible to both broaden the sources of supply and acquire the ability to manufacture components or completed weaponry on Chinese soil. Thus, one seeks more than mere prototypes or outright transfers of particular weapons systems. By acquiring the means of production itself— through the building of indigenous production facilities and the training of Chinese scientists and engineers to oversee such operations—military planners will be able to maintain their autonomy from external control.

It is in this context that the Sino-British jet engine agreement of late 1975 is most appropriately viewed.[80] Negotiations were initially undertaken as early as 1972, and proceeded more intensively during 1974 and 1975. The final agreement includes contractual obligations in three separate areas: (1) the initial supply of 50 supersonic Spey jet engines (the RB 168-25R, presently used in the British version of the Phantom F-4 and the Vought A-7 Corsair II close-support aircraft); (2) a license to manufacture these engines in China in a plant being built near Sian; and (3) the furnishing by Rolls Royce to the PRC of facilities and technical expertise for engine testing and maintenance. The ultimate result will be to advance Chinese jet propulsion technology by at least a half-dozen years, and provide China with the facilities to produce and maintain such engines on an independent basis by the early or mid-1980s. With Chinese personnel ultimately exercising full control over such plants, there is no possibility of undue (or unexpected) leverage being applied by the supplier state. Thus, within the foreseeable future, China's air force will for the first time possess an engine for fighter-aircraft whose capabilities and limitations are not intimately understood by their Soviet adversaries. Whatever the results of present negotiations between China and various Western European defense firms for modern weapons technology,[81] comparable arrangements will surely be a major objective of Chinese negotiators.

Given the extreme complexity and expense of such tasks, it should be little surprise that China during the 1950s turned so fully and unequivocally to the Soviet Union for precisely such assistance. The absence of alternative sources of supply and the willingness of Soviet policy makers to furnish such aid more than explains China's extreme dependence upon Soviet technology. Indeed, Chinese armaments production in the late 1970s is still based almost wholly on Soviet designs, some of it initially transferred more than two decades ago.[82] This fact illustrates the enormous difficulties of incorporating new designs and manufacturing facilities into a preexisting industrial structure of such size and consequence. It also indicates the irregularity with which defense modernization has been pursued over the past decade. Finally, it suggests the understandable reluctance of Chinese planners to commit themselves unequivocally to new plant investment unless one is persuaded that old weaponry simply no longer suffice.

While Chinese writings no longer discuss the full extent of Soviet support for China's defense modernization, both official and unofficial sources in the USSR continue to describe such aid. During the Korean war, according to one highly detailed article, "the cost—repaid by the Chinese side—of the arms equipment supplied to China by the U.S.S.R. . . . amounted to only 20 percent of the total value of the Soviet military credits." Moreover, it is further argued that more than half the military credits granted during the 1950-1955 period were used not for service in the Korean war, but as part of the PLA's overall modernization effort. Additional claims in this particular article seem wholly credible, and merit extensive quotation:

> Over the period 1950 through 1963, 71 enterprises of the military industry [out of more than 100 that had been pledged] were built in China with the participation of the Soviet Union. . . . The U.S.S.R. Government set aside for China from its own available stocks sufficient weapons and military-technical equipment to reequip 60 PLA infantry divisions. Equipment which was located in Port Arthur was also handed over to the PRC. At the same time, the Soviet Union gave China documentation for organizing the production of new models of practically all types of modern military equipment, and sent a large number of specialists there who gave assistance both in setting up the production of new types of military equipment and also mastering the armaments which the PLA military units had received. Thanks to Soviet military assistance the PRC was able, prior to 1960, to devote less than 10 percent of its budget to military purposes.[83]

The scope of such assistance may well be unprecedented in the history of alliances. This conclusion is even more apparent when one adds the very substantial assistance given by the Soviet Union to the Chinese nuclear program, aid that has been singularly instrumental to Chinese successes in the strategic weapons area.[84] There is little doubting Khrushchev's rather regretful conclusion that "all the modern weaponry in China's arsenal [in the early 1960s] . . . was Soviet-made or copied from samples and blueprints provided by our engineers, our research institutes. We'd given them tanks, artillery, rockets, aircraft, naval and infantry weapons. Virtually our entire defense industry had been at their disposal."[85] It hardly seems surprising that Mao's determination to break close ties with the Soviet Union was so strongly resisted by some of China's leading military officers.

Indeed, notwithstanding the total rupture of Sino-Soviet defense relations in the early 1960s, one should not rule out the possibility that newer Soviet weaponry might ultimately (if rather unconventionally) find its way into Chinese inventories. Several Japanese press reports bear directly on this question.[86] According to these sources, Egypt and China reached agreement in 1976 whereby China would receive a variety of Soviet armaments originally furnished to Cairo in exchange for spare parts and maintenance help for Egyptian MiG-17s and 21s damaged in the 1973 war. Some cash may also have been involved. The transaction supposedly included an unspecified number of MiG-23 aircraft, surface-to-air missiles, antitank weaponry, and T-62 tanks.

Assuming the veracity of such reports, intriguing possibilities become available to Chinese defense engineers, especially in fighter aircraft. A hybrid fighter-plane (tentatively designated the F-12) is purportedly being based on an improved model of the Spey engine, with numerous other components being drawn from the MiG-23. If more advanced Soviet technology in areas such as airframes could be utilized for such an aircraft, it could be easily adapted to preexisting defense industry facilities originally built by Soviet engineers. Comparable prospects would seemingly exist in other areas where such weaponry might be available. Should such efforts bear fruit, it would lend even greater support for arguments favoring continued development of indigenous defense industries, since their existence alone makes this possibility feasible.

Access to technology, therefore, exerts a singular influence on the pace and direction of China's military modernization. But this objective cannot be separated from the development process as a whole. Chinese writings now regularly discuss the ambitious goal of the "four modernizations" (agriculture, industry, national defense, and science and technology).

However, attention to the complex interrelationships among these four priority areas remains highly guarded. Rather than consider this topic in any detail, we will only indicate some of the more vital considerations that undoubtedly affect Chinese decision making.

If a single conclusion should be apparent from the discussion in this section, it concerns the specialization inherent in advanced defense technology. Some (though by no means all) of the needs generated by national defense requirements necessitate the investment of time, funds, and manpower that will be of only modest benefit to other industrial sectors. Thus, a key consideration in resource allocation is how to best integrate China's overall economic needs with the perceived imperatives of national security. This issue was less of a concern in the first two decades of Communist rule, and for several reasons. During the 1950s, the Soviet Union (as previously discussed) gave China access to an exceptional range of defense technologies, including the basic infrastructure for an entire modern defense industry. As the previously cited Soviet article noted,[87] it was through Soviet assistance that China was able to reduce vastly the percentage of state expenditure committed to national defense. Thus, while Chinese investment costs in machine-building and other defense-related industries were substantial, it was not nearly as dislocating as it would have been had Soviet aid not been available.

While the Soviet withdrawal of 1960 had a very pronounced effect on both civilian and military programs in China, it still did not fundamentally alter the institutional arrangements established in the 1950s. Thus, the defense plants already built with Soviet help or constructed outright by Soviet engineers could and did continue to produce weaponry deemed adequate for China's perceived national security needs. Moreover, by concentrating principally on security needs in terms of deterrence (i.e., the development of nuclear delivery systems and reliance on mass mobilization), China avoided the vexing decisions that would have been required if intermediate defense technology needs had been judged more pressing.

It is precisely such choices, however, that Chinese policy

makers have had to consider during the 1970s. As early as May 1971, articles in the Chinese press acknowledged that the highly specialized requirements of nuclear delivery systems and advanced weaponry more generally directly clashed with more basic investment needs in industrial development.[88] The posing of two alternatives ("electronics versus iron and steel") was somewhat disingenuous, in that Lin Piao was accused of overemphasizing the former over the latter. It would have been more accurate to admit that, insofar as Chinese defense needs focused very heavily on the nuclear weapons program, it tended to somewhat restrict the range of technologies (and the amount of investment) that a more expansive defense modernization program would have entailed.

By the mid-1970s, Chinese military planners were beginning to look far more candidly toward their vulnerabilities in "middle range" defense considerations. Attacks on Teng Hsiao-p'ing in 1976 focused on his supposed assertion that "fighting a modern war means fighting a war of steel."[89] Teng's opponents recognized that greater attention to a more differentiated national security agenda would unquestionably pose the issue of investment priorities in far more dramatic and consequential fashion than at any previous time.

This judgment has been amply borne out by the greatly heightened attention to military modernization apparent since the fall of 1976. To be sure, most analyses in the Chinese press (including those with military authorship) continue to assert that China's defense modernization still must follow overall improvements in economic construction. Yet a considerable number of articles now voice with increasing candor and explicitness a rather different argument.[90] In overall terms, they suggest that a militarily secure China in the 1980s cannot depend on the investment priorities that have heretofore been deemed adequate. As one particularly pointed article has stated:

> In any future war against aggression, if anyone still thinks it's possible to use broadswords against guided missiles . . . then he evidently is not prepared to possess all the weapons and means of fighting which the enemy has or may have. This is a foolish and even criminal attitude. . . .

Any future war against aggression will be a people's war under modern conditions. The suddenness of an outbreak of modern war, the complexity of coordinating ground, naval, and air operations, the extreme flexibility of combat units and the highly centralized, unified, planned, and flexible command structure—all these factors make it necessary for our army to have appropriate modern equipment.

For example . . . our armed forces must have an automatic computerized countdown, communications, and command system and rapid, motorized, modern transportation facilities. They must also be armed with conventional and strategic weapons so they can take quick and effective retaliatory action against any invading enemy. . . . Once [our armed forces] are armed with modern weapons, they will be like winged tigers and will become more invincible than ever.[91]

Quite clearly, the consequences of such altered priorities would be profound. Even assuming substantial growth in the Chinese economy as a whole, the structural implications of attending to such investment needs would fundamentally affect the orientation of the PRC's overall modernization program.

How are the potential conflicts between the expressed needs of defense planners and the more basic requirements of Chinese industrialization to be reconciled? It is still too early in this process to offer conclusive judgments. Nevertheless, several more refined policy arguments have already surfaced. One, voiced by numerous civilian decision makers, is to assert that defense needs, no matter how urgent they might appear, must still await sustained growth and improvements in basic industries and science and technology.[92] A second opinion, voiced in articles under military authorship, while acknowledging the importance of the development of the national economy as a whole, further asserts that the "defense industry . . . has considerable independence and initiative . . . [which] will inevitably continue to make new demands on other industries and on science and technology, thus motivating the development of the entire national economy."[93] A third viewpoint, aired by leaders charged with somehow reconciling such rival claims, offers the prospect of achieving simultaneous development: "Serious efforts should be made to implement the policy of integrating military with non-military enterprises and peacetime production with preparedness against war, and fully tap the potential of the machine-building and national defense industries."[94] All such arguments, however, are necessarily somewhat self-serving. They reflect the increasingly complex interactions between technology, economics, and national defense that will increasingly preoccupy Chinese policy makers in the coming decade, and for which no intermediate solutions or compromises presently seem discernible.

The Future of Chinese Military Power

This essay has necessarily been somewhat exploratory. While its findings must be considered tentative, certain judgments and conclusions seem inescapable. In overall terms, it is clear that Chinese power cannot be viewed from an abstract or mechanistic perspective. To be sure, we should be wary of considering Chinese behavior strictly according to categories suggested or implied by the Chinese themselves. Yet we should be even more guarded in depicting the conduct of China's leaders and armed forces as necessarily derived from Western practices. This is not to suggest that Chinese cultural peculiarities are so distinct as to defy comparison with foreign forms and traditions. Rather, the most appropriate view is to focus upon the institutional and historical sources of Chinese policies and programs, as they have interacted with the technological, strategic, and organizational imperatives affecting the acquisition of modern military power.

This brief review has tried to indicate that the impediments to substantial institutional change are perennial challenges, not momentary obstacles. The continuities in problems encountered by military leaders during nearly three decades in power are insufficiently recognized by many students of contemporary China. To be sure, the necessity for acquiring modern military power has never been the subject of serious challenge among China's leaders. Beyond a minimal consensus on this objective, however, substantial diversity in opinion and outcome has long been apparent: what to acquire, how much, how quickly, by what means, and for what purposes.

The continued maintenance of alternative forms of military organization and varied conceptions of military power amply bear out this judgment. Such diversity and complexity in military tasks remain a key element in the PLA's organizational ethic, as personally canonized by Mao. The ongoing dynamic between internal and external conceptions is not an organizational or political fiction: it epitomizes the needs and circumstances of both political and military leadership in China. In cumulative terms, it remains a major constraint on the exercise of Chinese power beyond the nation's borders. Under conditions where both traditions have long been impressively represented at the highest levels of political power, any major diminution in this duality of belief and organizational structure is likely to remain an exceedingly complicated and potentially conflictful task.

Within these two overall approaches to military power and policy, substantial variation of opinion and action is also apparent. As noted earlier, Chinese military forces have operated in a wide variety of physical settings and combat conditions that have required a substantial array of military skills and capabilities. Attention to these varied conflict situations has long necessitated highly distinct forms of organizational, economic, and technological effort. The potential incompatibilities in resource demands, institutional requirements, and military doctrines are no less evident today than they have been in previous decades. Indeed, under circumstances where defense modernization and security needs are being more openly addressed than at any time in the past two decades, such conflicts may simply become more acute. Close and continuing attention to where and how Chinese planners decide to concentrate their energies and available resources will unquestionably reveal much about dominant modes of thought and policy among those responsible for the security of China.

In addition, the particularly consequential nature of such decisions for resource allocation, technology acquisition, and basic industrialization remains a singular leadership consideration. Such issues emerged with unexpected suddenness in the Korean war, and they continue to preoccupy Chinese policy makers a quarter century later. Indeed, by their very nature they must remain vital tasks for the indefinite future. Alternative choices undeniably exist; moreover, they are inextricably linked to issues of threat perception, defense strategy, and the broader relationship between military capabilities and foreign policy. Such complex interrelationships can only be assessed and understood by recognizing the degree to which military policy interacts with more basic economic, scientific, and technological issues.

In a broader sense, however, Chinese policy makers must soon confront an even larger issue: the overall purposes to which military power can and should be put. For much of the past three decades, Chinese security policy could be legitimately described as largely reactive to external pressures and imperatives. Over the next ten years, China will begin acquiring increasingly consequential military capabilities. Such power will offer the potential for a defense posture that is far more explicitly oriented beyond China's national boundaries.

Those responsible for Chinese security are thus approaching a crossroads. The invidious label of superpower, as used by the Chinese, reflects a judgment about both the capabilities and intent of states. As Teng Hsiao-p'ing argued at the United Nations in April 1974: "A superpower is an imperialist country which everywhere subjects other countries to its aggression, interference, control, subversion, or plunder, and strives for world hegemony. . . . China would turn into a superpower if she too should play the tyrant in the world, and everywhere subject others to . . . bullying, aggression, and exploitation."[95] Thus it is not the military and political might of major powers per se that is evil, but the purposes to which such strength can be put. The pledge of Teng and others that "China is not a superpower, nor will she ever seek to be one" is credible and legitimate only to the extent that conscious restraints in the exercise of Chinese power are upheld.

If past history is any guide, however, the new opportunities afforded states acquiring substantial military capabilities have almost always been pursued, in whole or in part. China is now embarked on an ambitious effort to move to the front ranks of the world's powers by the year 2000. Leaders in Peking may not have long to wait to see whether the capacity to use force beyond the nation's borders—and hence, like most great powers, pursue a more expansive conception of national security—creates inexorable pressures to move in that direction. These are clearly considerations that Chinese elites will not be able to ignore, as the recent Vietnam case so effectively illustrates.

Thus, while the specifics of the Chinese policy process remain difficult to penetrate or observe, this does not eliminate the possibility of meaningful analysis and prediction. The continuities in the national defense agenda in China and the comparability it suggests with security policy in other states are cause for encouragement on the prospects for further inquiry. Under conditions where China's leaders are increasingly explicit and candid about precisely such questions, the tasks for further study seem obvious, relevant, and feasible.

Notes

1. Mao Tse-tung, "Address to the Preparatory Committee of the New Political Consultative Congress," June 15, 1949, in *Selected Works of Mao Tse-tung*, vol. 4 (Peking: Foreign Languages Press, 1961), pp. 406-407.

2. Mao Tse-tung, "The Chinese People Have Stood Up!" September 21, 1949, in ibid., vol. 5 (Peking: Foreign Languages Press, 1977), p. 18.

3. Allen S. Whiting, "The Use of Force in Foreign Policy by the People's Republic of China," *The Annals* 402 (July 1972):55-66, and *The Chinese Calculus of Deterrence: India and Indochina* (Ann Arbor: The University of Michigan Press, 1975), especially chapters 7-8. See also Commander Bruce Swanson, "The P.R.C. Navy: Coastal Defense or Blue Water?" *U.S. Naval Institute Proceedings* 102 (May 1976):89-90. For discussion and analysis of China's attack on Vietnam in early 1979 (the eleventh and most recent of these episodes), see Drew Middleton's reports in the *New York Times* during February and March 1979.

4. Both figures are taken from *The Military Balance, 1977-1978* (London: International Institute for Strategic Studies, 1977), pp. 53-54.

5. According to U.S. government data, between 1967 and 1976 China transferred approximately $2.6 billion (in 1975 constant dollars) in arms to Third World states, thereby making the PRC the world's fifth leading arms exporter. *World Military Expenditures and Arms Transfers, 1967-1976* (Washington: U.S. Arms Control and Disarmament Agency, 1978), p. 126.

6. Such a perspective remains the dominant orientation among students of Chinese security, no doubt in large part due to the informational limitations on decision making in the PRC. For representative examples, see Davis B. Bobrow, "Peking's Military Calculus," *World Politics* (January 1964):287-301; and Arthur Huck, *The Security of China: Chinese Approaches to Problems of War and Strategy* (New York: Columbia University Press, 1970). In a somewhat different vein, see John Gittings, *The World and China, 1922-1972* (New York: Harper and Row, 1974).

7. On these issues, see in particular W. W. Whitson, *Chinese Military and Political Leaders and the Distribution of Power in China, 1956-1971*, R-1091-DOS/ARPA (Santa Monica, Calif.: Rand Corporation, 1973); and George C. S. Sung, *A Biographical Approach to Chinese Political Analysis*, R-1665-ARPA (Santa Monica, Calif.: Rand Corporation, 1975).

8. To focus on a limited number of issues, however, presupposes a higher degree of confidence in the present state of knowledge and understanding than is probably warranted. Thus, our analysis should be considered partial and tentative, and subject to modification in the course of additional inquiry.

9. Unless otherwise indicated, references to the PLA are meant to encompass the entire spectrum of Chinese military organizations: e.g., ground forces, air and naval units, missile forces, etc.

10. I have discussed this question in "The Study of Chinese Military Politics: Toward a Framework for Analysis," in *Political-Military Systems: Comparative Perspectives*, ed. Catherine M. Kelleher (Beverly Hills: Sage Publications, 1974), in particular pp. 253-59.

11. Harvey Nelsen, *The Chinese Military System* (Boulder, Colorado: Westview Press, 1977), especially chapters 3-7.

12. For a review of the PLA's continuing involvements in political and economic management, see *The PLA and China's Nation-Building*, ed. Michael Ying-mao Kau (White Plains, N.Y.: Interna-

tional Arts and Sciences Press, 1973). Other scholars, however, dispute the view that the PLA remains so heavily committed to such tasks. See, for example, Nelsen, *The Chinese Military System*, pp. 143-45.

13. See, for example, "Turn the Army into a Working Force" (February 8, 1949), in *Selected Military Writings of Mao Tse-tung* (Peking: Foreign Languages Press, 1963), pp. 391-93; and "Instructions on the Army's Participation in Production and Construction Work in 1950," (December 5, 1949), in *Mao Tse-tung Ssu-hsiang Wan-sui!* [Long Live Mao Tse-tung Thought] (n.p., 1969), pp. 1-4.

14. Ibid.

15. Sung, *A Biographical Approach*, pp. 77, 120. Sung's category of "conventional soldiers" includes those affiliated with three military sectors: armor, infantry, and public security. No other sector, military or nonmilitary, has ever comprised more than fourteen percent; for the last year of his data (1973), the second highest percentage of representation is less than eight percent.

16. For a preliminary assessment of various trends in Chinese defense policy since the death of Mao, see my articles, "The Logic of Chinese Military Strategy," *Bulletin of the Atomic Scientists* (January 1979):22-33, and "Rebuilding China's Great Wall," forthcoming.

17. Considerations of space preclude a more detailed discussion of this topic and of the evidence used to deduce closer involvement with security issues. For a careful assessment on what is known about this overall division of labor, consult Nelsen, *The Chinese Military System*, especially chapter 3.

18. For references to the principal documents, consult Michael Yahuda, "Kremlinology and the Chinese Strategic Debate," *The China Quarterly*, no. 49 (January-March 1972):32-75.

19. Beyond a general consensus on this point, there is pronounced disagreement among those who have studied these events. For the respective views, consult Uri Ra'anan, "Peking's Foreign Policy Debate, 1965-1966," in *China in Crisis*, vol. 2, ed. Ping-ti Ho and Tang Tsou (Chicago: University of Chicago Press, 1968), pp. 23-71; Donald Zagoria, "The Strategic Debate in Peking," in ibid., pp. 237-68; and Yahuda, "Kremlinology."

20. Ra'anan in particular maintains this view—leading him to suggest that in this internal Chinese debate the United States should have been "rooting for Mao."

21. Peking: Foreign Languages Press, 1965.

22. Whiting, *The Chinese Calculus of Deterrence*, pp. 186-87.

23. For example, PLA MiG-17s and MiG-19s were based in North Vietnam as early as August 1964. Three new airfields in southern China immediately proximate to the DRV borders were under construction by October. And in January, PRC and DRV jets were engaged in joint air exercises along the Sino-Vietnamese border. Ibid., pp. 175-77.

24. Yahuda, "Kremlinology," pp. 41-42.

25. This section draws heavily from Harry Harding and Melvin Gurtov, *The Purge of Lo Jui-ch'ing: The Politics of Chinese Strategic Planning*, R-548-PR (Santa Monica, Calif.: Rand Corporation, 1971).

26. For the best account of the PLA's involvement in domestic politics leading up to the Cultural Revolution, see Ellis Joffe, "The Chinese Army Under Lin Piao: Prelude to Political Intervention," in *China: Management of a Revolutionary Society*, ed. John M. H. Lindbeck (Seattle: University of Washington Press, 1971), pp. 343-74.

27. The Work Group of the CCP Central Committee, "Report of the Problem of Lo Jui-ch'ing's Mistakes," *Chung-kung Nien-pao* [Yearbook on Chinese Communism] (Taipei: Institute of International Relations, 1970), reprinted in Kau, *The PLA and China's Nation-Building*, p. 298. The above internal document was acquired by Chinese Nationalist intelligence and subsequently disseminated publicly.

28. For one such accusation, see Feng Chu-hsia and Chi Chang-hung, "What Is Behind His Trumpeting About 'A Matter of Secondary Importance'?" *Kwangming Jih-pao* [Enlightenment Daily], September 12, 1974.

29. See the article by the Hsinhua correspondent in *Jen-min Jih-pao*, October 14, 1977, p. 1.

30. "The Historical Experience of Three Ups and Three Downs in Military Training," *Chieh-fang-chün Pao*, January 24, 1978. This article was reprinted the following day in *Jen-min Jih-pao*. In view of what is known about the PLA's turbulent politics since the mid-1960s, the article's principal assertions do not seem implausible.

31. Ibid.

32. Ibid.

33. All the above quotes are drawn from Su Yü, "Great Victory for Chairman Mao's Guideline on War," *Jen-min Jih-pao*, August 6, 1977. It is translated in part in *Peking Review*, no. 34 (August 19, 1977):6-15.

34. These quotes are drawn from a decision of the CCP Military Commission on "Strengthening the Army's Education and Training," as reported in *Chieh-fang-chün Pao*, Peking Domestic Service [FBIS] *Daily Report-People's Republic of China*, May 15, 1978, pp. E9-10; and *Jen-min Jih-pao*, May 14, 1978, p. 1, in ibid., pp. E10-12.

35. See in particular "Problems of Strategy in China's Revolutionary War," (December 1936), in *Selected Military Writings of Mao Tse-tung*, pp. 75-150.

36. On Chinese operations in the Korean War, see Alexander George, *The Chinese Communist Army in Action: The Korean War and its Aftermath* (New York: Columbia University Press, 1967), especially chapters 1, 9. On the Chinese intervention more broadly, see Allen Whiting, *China Crosses the Yalu* (Stanford: Stanford University Press, 1968); and J. H. Kalicki, *The Pattern of Sino-American Crises* (London: Cambridge University Press, 1975), chapter 2.

37. On the Sino-Indian war, consult Whiting, *The Chinese Calculus of Deterrence*, especially chapters 4-5.

38. This essay will not dwell extensively on actual military tactics; it is only intended to convey the broader outlines of the forms of conflict in which PLA forces can engage. For an extremely detailed discussion of Chinese tactical doctrine for infantry forces, see Defense Intelligence Agency, *Handbook on the Chinese Armed Forces*, DDI-2680-32-76 (Washington, D.C.: Department of Defense, 1976), chapter 4.

39. George, *The Chinese Communist Army in Action*, pp. 161, 163, 188-89.

40. See, in particular, "On the Great Victory in the War to Resist U.S. Aggression and Aid Korea and our Future Tasks," September 12, 1953, in *Selected Works of Mao Tse-tung*, vol. 5, pp. 115-120. Some of Mao's glowing accounts of Chinese battlefield successes suggest that he had not been wholly apprised about the enormity of Chinese losses and reversals in the spring of 1951.

41. Ibid., p. 115.

42. For two recent articles that place great stress on the close coordination of military action to political objectives during the Korean conflict, see the "Reference Information" column in *Jen-min Jih-pao*, May 10, 1977, p. 2; and the analysis by Lu Tan-sheng in *Kwang-ming Jih-pao*, June 3, 1977, p. 2.

43. Mao Tse-tung, "Speech at First Plenum of Ninth CCP Central Committee" (April 1969), in *Joint Publications Research Service* (JPRS), no. 50,564, as cited in Harding and Gurtov, *The Purge of Lo Jui-ch'ing*, p. 53n.

44. For graphic evidence of these considerations, compare George, *The Chinese Communist Army in Action*, chapter 9, with recent assessments of Chinese ground force preparedness, in particular, Nelsen, *The Chinese Military System*, chapter 4; Drew Middleton, "Visit to China's Forces: Big but Poor in New Arms," *New York Times*, December 1, 1976, and "Chinese Military Regards Manpower as No. 1 Asset," ibid., December 2, 1976; and Russell Spurr, "China's Defense: Men Against Machines," *Far Eastern Economic Review*, January 28, 1977, pp. 24-30.

45. Whiting, *The Chinese Calculus of Deterrence*, chapter 3.

46. Very little consensus exists on the specific circumstances surrounding the March 1969 border clashes. The preponderant opinion, reflecting the greater availability of information from Soviet sources, is offered in Thomas W. Robinson, "The Sino-Soviet Border Dispute: Background, Development and the March 1969 Clashes," *American Political Science Review* (December 1972), especially pp. 1187-90. Since the publication of Robinson's study, Chinese sources have provided a far more detailed account of these events than had previously been offered. It is presented by Neville Maxwell in "The Chinese Account of the 1969 Fighting at Chenpao," *The China Quarterly* 56 (October-December 1973):730-39.

47. Maxwell, for example, notes that Chinese forces have held Chenpao Island continuously since March 17, 1969. Yet the more significant result of these limited exchanges has been the continuation and enhancement of enormous military capabilities on both sides of the border. *The China Quarterly* 56 (October-December 1973):738.

48. For example, these conceptions seem supported by what is known about the involvements of navy units in civilian tasks such as protection of fishing fleets. Swanson, "The P.R.C. Navy," pp. 99-103; and Lieutenant David G. Muller, Jr., "The Politics of the Chinese People's Republic Navy," *Naval War College Review* (Spring 1976):32-51. They are given further credence in discussions found in the *Kung-tso T'ung hsün* [Work Bulletin], a top-secret periodical of the PLA General Political Department for which selected issues from 1961 are available. See, for example, "We Must Do Good Substantial Work in Building Up the Militia," in *Work Bulletin* 21 (May 26, 1961), as translated in *The Politics of the Chinese Red Army*, ed. J. Chester

Cheng (Stanford: The Hoover Institution, 1966), pp. 559-564.

49. Whiting, *The Chinese Calculus of Deterrence*, chapter 3.

50. John Gittings, *The Role of the Chinese Army* (New York: Oxford University Press, 1967), pp. 40-44; Kalicki, *The Pattern of Sino-American Crises*, chapter 6; Thomas Stolper, "The Taiwan Strait Crisis of 1954-55" (University of Michigan, unpublished manuscript, February 1975); Jonathan D. Pollack, "Perception and Action in Chinese Foreign Policy: The Quemoy Decision" (Ph.D. diss., University of Michigan, 1976), chapters 4-6.

51. See, for example, Shih Ti-tsu, "South China Sea Islands, Chinese Territory Since Ancient Times," *Kwangming Jih-pao*, November 19, 1975, reprinted in *Peking Review* 50 (December 12, 1975):10-15.

52. For a discriminating discussion of the growth of naval capabilities and of the expanding mission structure that has accompanied it, see David G. Muller, Jr., "The Missions of the PRC Navy," *U.S. Naval Institute Proceedings* 103 (November 1977):47-52.

53. These operations are described in Swanson, "The P.R.C. Navy." For a recent Chinese account of these events, see the article by naval officer Wu Kung-chi in Peking in Mandarin to Taiwan, August 1, 1977, in FBIS *Daily Report-P.R.C.*, August 4, 1977, pp. E9-10.

54. The details of these exercises remain rather sketchy. According to congressional testimony by General George Brown, former chairman of the Joint Chiefs of Staff, these training exercises lasted a full week. They included the use of "attack aircraft . . . in bombing and strafing activity in support of parachute assaults with the entire force being protected by interceptors. Although the exercise was elementary by U.S. standards, it did show increased sophistication in the use of tactical air power in the P.R.C." Interestingly, Chinese broadcasts have described such exercises in comparable terms. See the August 11, 1977, article in *Chieh-fang-chün Pao*, as cited in FBIS *Daily Report-P.R.C.*, August 12, 1977, pp. E4-5; and Peking Domestic Service in Mandarin, October 3, 1977, in ibid., October 5, 1977, p. E4. See General Brown's *Statement to the Congress on the Defense Posture of the United States for Fiscal Year 1978*, January 20, 1977, p. 84, and *Statement for Fiscal Year 1979*, January 20, 1978, pp. 83,116.

55. See, for example, the April 28 comments of Wu Hsiu-chuan, a PLA deputy chief of staff, to a visiting group of Japanese military experts, as excerpted in Hong Kong *Ta Kung Pao*, May 1, 1978, in FBIS *Daily Report-P.R.C.*, May 5, 1978, pp. N1-2.

56. For the most detailed and forceful articles advocating expanded

naval development, see *Jen-min Jih-pao*, March 15, 1977, p. 2, and June 24, 1977, p. 3.

57. This is now acknowledged in Soviet sources, who further report that entire air defense wings protecting cities in Manchuria were manned by Soviet personnel. See, for example, O. B. Borisov and B. T. Koloskov, *Sino-Soviet Relations, 1945-1973* (Moscow: Progress Publishers, 1975), p.51.

58. R. M. Bueschel, *Chinese Communist Air Power* (New York and Washington: Frederick A. Praeger, 1968), pp. 20-28.

59. Pollack, *Perception and Action in Chinese Foreign Policy*, especially chapter 6.

60. For a recent review of these capabilities, see Russell Spurr, "Ivan's Arms Around Manchuria," *Far Eastern Economic Review* (January 28, 1977):26.

61. Statement of Morton I. Abramowitz, deputy assistant secretary of defense, East Asia and Pacific affairs, April 6, 1976, in U.S., Congress, House Subcommittee on Future Foreign Policy Research and Development of the Committee on International Relations, *United States-Soviet Union-China: The Great Power Triangle*, hearings, 1976, p. 184.

62. William Beecher, "Shift in Strategy by Peking Seen," *New York Times*, July 25, 1972, pp. 1, 14.

63. For the fullest first-hand accounts, see Drew Middleton's three-part series in the *New York Times*, December 1-3, 1976; and Edward N. Luttwak, "Chinese Strategic Security After Mao," *Jerusalem Journal of International Relations* (Summer 1977):97-111.

64. This is obviously much too complex a topic for extended discussion in this essay. For an overall assessment, see my "Rebuilding China's Great Wall," *Bulletin of the Atomic Scientists*, forthcoming.

65. See Middleton, *New York Times*, and Luttwak, "Chinese Strategic Security."

66. I have undertaken a much more extended analysis in "China as a Nuclear Power," in *Asia's Nuclear Future*, ed. William H. Overholt (Boulder, Colo.: Westview Press, 1977), pp. 239-70.

67. Su Yü, "Great Victory for Chairman Mao's Guideline on War," p. 15.

68. For some recent examples, see the following articles in *Jen-min Jih-pao*: Hu Hsüeh, "Get Rid of Blind Belief in Nuclear Weapons," May 13, 1977, p. 6; Chi Chuan, "Long Live the Spirit of Millet Plus Rifles," June 3, 1977, p. 5; Chieh Chan, "The Atom Bomb is a paper Tiger," June 21, 1977, p. 5; and Hsia Li, "One Thing Surpasses

Another," July 26, 1977, p. 5.

69. Though we will touch upon this issue briefly in the next section of this essay, it is too complicated an issue for extended consideration in this context. For further discussion, see Pollack, "The Logic of Chinese Military Strategy," and "Rebuilding China's Great Wall."

70. In the first known instance where Mao explicitly advocated the development of nuclear weaponry, he argued that "if we are not to be bullied in the present day world, we cannot do without the bomb." Mao Tse-tung, "On the Ten Major Relationships" (April 25, 1956), in *Peking Review*, no. 1 (January 1, 1977):13.

71. I am grateful to John Wilson Lewis for calling this fact to my attention. For evidence bearing out his contention, see *Handbook on the Chinese Armed Forces*, pp. 3-12, 3-15.

72. Lin Piao, "Basic Differences Between the Proletarian and Bourgeois Military Lines," *Peking Review*, no. 48 (November 24, 1967), pp. 11-16.

73. "Give Full Play to the Great Role of the Militiamen in Their Hundreds of Millions," *Chieh-fang-chün Pao* editorial, June 19, 1977, reprinted on the same date in *Jen-min Jih-pao*, p. 2.

74. Statement of George Bush (then director of Central Intelligence), May 27, 1976, in U.S., Congress, Subcommittee on Priorities and Economy in Government of the Joint Economic Committee, *Allocation of Resources in the Soviet Union and China—1976*, hearings, pt. 2, 1976, p. 31.

75. *World Military Expenditures and Arms Transfers, 1967-1976*, p. 39. Although comparisons are not wholly apt, the 1976 data on the other major military spenders offers some perspective on this figure: USSR—$121 billion; United States—$87 billion; Federal Republic of Germany—$15 billion; and France—$14 billion. Ibid., p. 6.

76. Bush, *Allocation of Resources*, p. 31.

77. For recent Chinese sources that show an explicit awareness of central dimensions of this modernization process, see "March Toward the Modernization of Science and Technology for National Defense," *Chieh-fang-chün Pao* editorial, September 24, 1977, in *Jen-min Jih-pao*, September 25, 1977, p. 3; "Integration of 'Millet Plus Rifles' With Modernization" (article written by the theoretical group of the National Defense Science and Technology Commission), in Peking Domestic Service in Mandarin, January 20, 1978, in FBIS *Daily Report-P.R.C.*, January 23, 1978, pp. E1-6; and *Chieh-fang-chün Pao* newsletter, in Peking Domestic Service in Mandarin,

March 22, 1978, in ibid., March 24, 1978, pp. E12-13.

78. Hans Heymann, Jr., *China's Approach to Technology Acquisition: Part I—The Aircraft Industry*, R-1573-ARPA (Santa Monica, Calif.: Rand Corporation, 1975), pp. 18, 23-24. Only two prototypes of the TU-16, a mainstay of China's nuclear delivery systems, were originally available to the PRC; they had been transferred to China in 1960.

79. Nikolai Cherikov, "The Shenyang F-9 Combat Aircraft," *International Defense Review* (October 1976):714-16.

80. This paragraph derives principally from "Breaking with the Past," *Far Eastern Economic Review* (December 26, 1975):9; U.S., Congress, House, Subcommittee on International Relations, *Export Licensing of Advanced Technology: A Review*, hearings, April 12, 1976, pp. 7-8, 26-27; and "U.K. Assisting China in Spey Production Plan," *Aviation Week and Space Technology*, July 12, 1976, p. 16.

81. See, for example, the following articles in the *New York Times*: Drew Middleton, "Chinese Showing New Interest in Buying Western Arms to Help Modernize Army," February 28, 1977; Fox Butterfield, "Peking Shows Interest in Purchasing Western Arms," October 23, 1977; Drew Middleton, "China Looking to Western Europe for Arms Supplies," April 14, 1978; and "Peking Said to Buy Missiles in France," May 3, 1978.

82. Abramowitz, *United States-Soviet Union-China*; Drew Middleton, "What the Chinese Forces Lack: Most Types of Modern Weapons," *New York Times*, June 24, 1977; Brown, *Statement to the Congress*.

83. All the above citations are taken from O. Ivanov, "Peking's Falsifiers of the History of Soviet-Chinese Relations," *Mirovaya Ekonomika i Mezhdunarodnyye Otnosheniya* 12 (November 19, 1975), trans. in FBIS *Daily Report-Soviet Union*, January 14, 1976, pp. C8-9.

84. This issue is discussed much more extensively in Pollack, "China as a Nuclear Power," pp. 38-41.

85. Nikita Khrushchev, *Khrushchev Remembers—The Final Testament*, trans. Strobe Talbott (Boston: Little, Brown, 1974), p.269.

86. See, the Kyodo report from Peking, November 14, 1977, in FBIS *Daily Report-P.R.C.*, November 15, 1977, p. E1; and a second Kyodo report from Tokyo (allegedly based on information from Japan Defense Agency sources), January 19, 1978, in ibid., January 20, 1978.

87. Ivanov, "Peking's Falsifiers," p. C9.

88. See, "Develop China's Iron and Steel Industry Under the

Guidance of Mao Tse-tung Thought," *Jen-min Jih-pao*, May 12, 1971, p. 3; and "A Criticism of the Theory of Making the Electronics Industry the Center," *Jen-min Jih-pao*, August 12, 1971.

89. See, for example, Peking Domestic Service in Mandarin, August 4, 1976, in FBIS *Daily Report-P.R.C.*, August 10, 1976, pp. E4-5; and the article by Shen Ping in *Hung-ch'i*, no. 8 (August 1976), in ibid., August 24, 1976, pp. E1-7.

90. For a more detailed examination of these arguments, see Pollack, "Rebuilding China's Great Wall."

91. "Integration of 'Millet Plus Rifles' With Modernization," FBIS *Daily Report-P.R.C.*, January 23, 1978, pp. E3-4.

92. See, for example, Li Hsien-nien's speech to the National Conference on Learning from Ta-ch'ing in Industry, April 20, 1977, in *Jen-min Jih-pao*, April 23, 1977.

93. "The Strategic Policy of Strengthening Defense Construction," *Kwang-ming Jih-pao*, January 20, 1977, p. 2. The authors of the article are identified as the theoretical group of the National Defense Industry Office.

94. The citation is from Hua Kuo-feng's report on the work of the government delivered at the first session of the Fifth National People's Congress, February 26, 1978, in *Peking Review*, no. 10 (March 10, 1978), p. 23.

95. Teng Hsiao-p'ing, speech at special session of U.N. General Assembly, April 9, 1974, in *Peking Review*, supplement to no. 15 (April 12, 1974):5.

4
India's Military Power and Policy

Onkar Marwah

Introduction

Over the past thirty years, India has slowly but steadily built up one of the world's largest armed forces establishments from the minimal forces existing in 1947. The country then possessed an army of 300,000 soldiers, an air force of two fighter and one transport squadrons, and a navy comprising four sloops, two frigates, and some harbor defense craft. The capacity to manufacture lethal armaments was almost nonexistent. At independence, the military officer corps consisted largely of noncommissioned officers, captains, and majors, and the country's new leadership had had little experience with the role or use of armed forces as instruments of state policy.

By 1979, India had acquired the world's third largest standing army, fifth largest air force, and eighth largest navy.[1] Its domestic armament industry was the biggest among Third World noncommunist states in value, volume, diversity of manufacture, and research and development facilities.[2] The country's military officer corps numbered 30,000-40,000, with substantial numbers among them trained for staff level duties. The Indian leadership had also absorbed lessons from four substantial external wars and one continuing internal war in northeast India. In all of those conflicts the country was engaged in negative interactions with one or another of the world's major military powers.

Through its recent nuclear and space activities, India has exhibited an ability—and some suggest the intent—to acquire strategic weapons and a delivery system at some point in the future.[3] The world's tenth largest industrial base and third largest supply of skilled and technical manpower have been supporting the military effort. Unique among Third World states, the increases in Indian military capability have been overseen throughout the past three decades by civilian regimes which have been sensitive to the functions of the military but which have allowed the latter no role in political decision making.

The preceding (and probably continuing) growth in Indian military capacity has been derived from a combination of circumstances. Some of these have been the result of deliberate national policy decision, some have been responses to the actions of other states, and some have been a function of the realities of the Indian political environment. In the first category were decisions pertaining to the international political role sought by India in the postwar period. Most importantly, these included the decision to avoid direct participation in military alliances while simultaneously seeking to enlarge nonmilitary interaction with both the liberal Western and socialist blocs of states. The second category of influences encompassed the actions of other states that were perceived negatively by India. The principle episodes included: the U.S.-Pakistani mutual security pact of 1954 and the consequent free supply of $2 billion worth of arms to Pakistan; the forceful manner in which China established its claim in 1962 to territories in dispute with India; the implications of the increasing great-power naval deployments in the seas around India; and the indirect effects from the flood of modern weapons being supplied to states around the Persian Gulf. To be included in this category in the future will be problems arising from the probable supply of Western arms and technology to China. The third category of influences has been the result of realities of the Indian domestic environment that make the country vulnerable to internal or external attack. The most important factors were the nation's continuing poverty, its burgeoning population, and the fractiousness of its multi-ethnic community of peoples.

This study explores the circumstances that have conditioned Indian national threat perceptions and have led to the steady increases in India's military power over the past thirty years. It will review the interaction of those circumstances and threat perceptions over time and within the geopolitical contexts relevant to the country. It will consider likely alterations in the geostrategic environments that those circumstances and threats create for India in the future. And finally, the study will assess Indian policy choices with respect to conventional and nuclear weapons acquisition in the succeeding years.

The major proposition of the study concerns India's long-term and "semipermanent" security anxieties. These arise from the mixture of conflict and conformity by which the Indian political and economic system relates to forms of state organization inspired by the world's two leading ideological systems, the socialist and the Western liberal. Being neither the one nor the other in total measure, India perceives itself as under pressure from the proponents of both ideological systems. The threat in this instance is internal as well as external, for numerous adherents to both ideologies exist within the country and could call for outside support in a period of national strife. The examples of Chile, Angola, Afghanistan, and Ethiopia suggest that Third World regimes can be undermined in a variety of ways. Though Indian policymakers remain reasonably confident of the strength and resiliency of their state system, anxiety lingers that events beyond their control, devolving from the cross-cutting global objectives and shifting strategic needs of the superpowers, could lead to unexpected pressures on India. Examples of such possibilities include permanent superpower naval deployments in the Indian Ocean with the consequent incentive to intervene in littoral-state affairs; change and counterchange in U.S.-Soviet relations with China; and unpredictable moves in major power arms transfer policies in relation to states within or bordering on South Asia.

A second proposition views India's short-term and hence "temporary" security needs as a function of unresolved historical disputes over the extent of Indian territory. The threats in these instances are external and physical. They have also been specific in terms of location: from China in relation to adjustments over the northern boundaries, and from Pakistan in relation to the ownership of Kashmir. The nature and determination of the threats in these cases probably lie somewhere between the two extremes of either acceding fully to Chinese and Pakistani territorial claims or rejecting them completely. For the moment, and despite an effort to normalize relations with both China and Pakistan, India has not modified its claim to territories wrested away by China in 1962, or the portion of Kashmir held by Pakistan since 1948.

The third proposition of this study refers to India's most "permanent" security threats. In both the short and long term, these are generated by the conditions of poverty, inequity, lack of domestic cohesion, and the expectations raised by the very processes of development and nation-building activity in which India will remain engaged for many decades. The threats in these instances are generic and internal. Depending upon the context, the challenges to state power could be physical or otherwise. They could derive from ethnicity-based secessionism, class-inspired peasant revolt, or (as during the recent Emergency) arise as a form of institutionalized civil and political conflict between factions of the ruling elites. Again, depending upon the situation, the internal threats could become externalized as powerful states received incentives for covert or overt interventions in the Indian arena.

To the extent that the preceding categories of security problems reflect the need for military resolution, Indian leaders have made decisions providing the country with a broad spectrum of technological and economic capacities linked to the nation's defense. Isolation from military blocs has inevitably demanded that the country be able as far as possible to establish its freedom from weapons-supplier states and their shifting policies. The military clash with China—which soon thereafter proceeded to build nuclear weapons—has meant that India will not easily surrender its own nuclear weapons option.

In a period of fluid relations between the United States, the Soviet Union, and China, the continued weapons buildup of the USSR and China, and their impressive abilities to undertake fundamental shifts in global strategy or messianic interventions abroad, Indian leaders see their own military effort as a necessary precondition to enhancing in a very broad sense their nation's security. The objective gains sought are varied and diverse. At one level, this effort is intended to insulate the country and its internal network of political and economic choices from involuntary change at the behest of external or internal adversaries. It is further designed to raise the stakes for external powers engaged in activities that complicate India's regional or subcontinental security goals. Finally, India's military programs seek to establish national self-images for the state commensurate with its size, numbers, resources, potential, and presumed role in the international political system.

The pursuit of a relatively independent conventional-weapons production and military power base and a possible strategic weapons system in the future is seen by Indian leadership as serving additional objectives in the subcontinent and in Asia as a whole. These include the neutralizing of Pakistan's conventional military capability to the point where states beyond the subcontinent would find it increasingly cost-ineffective to encourage or seek the creation of a military balance between India and Pakistan. They further involve matching (with a margin of safety in their own favor) China's conventional military forces in the Tibet and Sinkiang regions, and providing the basis for future parity with China's strategic weapons capacity. Finally, the weapons and military personnel are intended to compensate for the possibility that a large segment of the arms being supplied to states in the Persian Gulf end up in Pakistan.

It is as easy to overestimate the growth in Indian military power as it has sometimes been to underestimate it in the past. The country remains a bundle of contradictions: it possesses an industrial economy of major proportions and sophistication and at the same time contains large masses of poor people; it possesses large numbers of trained and educated people along with the world's largest bloc of illiterates; it retains the ability to provide internally for 95 percent of its ongoing requirements of industrial machinery and agricultural products yet struggles for the means—being hurt in the attempt—to acquire the remaining 5 percent from abroad. Notwithstanding such contradictions, this study does not envisage any sudden or dramatic departure from the prevailing policy framework. It is assumed that a pacifist leadership will not replace the present one in India and that the existing broad context of foreign and domestic policies will not be subject to fundamental redirection by Indian leaders. Second, it is believed that the country's gross economic growth rates will not fall below 3 to 4 percent per annum. Third, it is assumed that a wide-ranging militant revolt by a mobilized landless peasantry, which comprises over 40 percent of the national population, will not occur. Finally, it is assumed that intangibles of national behavior, such as will, morale, and resolve, will continue to remain relatively constant.

Irrespective of shortfalls, then, India is to be viewed as a large state with significant man-made and natural resources available to it along a variety of dimensions. On the Indian Ocean littoral, it is the biggest state in size, population, and skilled personnel, and has the most minerals (save oil), industrial capacity, and military power. Its capacity to build upon what now exists, therefore, is likely to remain greater than that of any individual state on the littoral. This assessment holds despite the vast arms flow into the Persian Gulf area, the petrodollar-financed economic and military activity in that region, or indeed the Gulf-oriented focus of great power interests. It is also likely that, being based progressively on a domestic industrial effort, India's military power will continue to maintain an edge over that of the other states in the Indian Ocean region. Under special conditions, it could outdistance the capacity of its putative rivals.

To consider the preceding assertions, we need to assess India's overall security record to date along various key dimensions. These are (1) the security perceptions and learning experiences of India's governing elites and the degree and manner in which they have varied over time; (2) the political and economic frameworks within which national elites moved to satisfy, over time, the varying needs of development and defense; and (3) the impact in regional and extraregional contexts of India's political and military choices.

The Security Experience and Perceptions of India's Governing Elites

During World War II India, then undivided, had the dubious distinction of fielding the largest volunteer land army in the world. Comprised of over 2.5 million men, Indian soldiers formed the biggest national element of allied forces in North

Africa, the Middle East, and the mainland of Southeast Asia. Indian army formations were heavily employed in the battles of El Alamein, the Cassino heights, Burma, Malaya, and Indo-China. Elite groups from within the Indian army—such as the Second Armored, Fourth Sikh, and Fifth Gurkha divisions—emerged as the most decorated units of the (then) "British Empire Forces." After the war, Indian troops performed garrison duty along the northern Indian Ocean littoral, in the Persian Gulf region, and in Southeast Asia, including the Indonesian archipelago. An Indian served on the Far Eastern War Crimes Tribunal in Tokyo, as did others in various postwar resettlement commissions, prisoner exchanges, and the like.[4]

Notwithstanding their wide involvement during and in the immediate aftermath of World War II, the (British) Indian government instituted a policy of swift postwar demobilization of the country's armed forces. By 1947, when the subcontinent gained political independence from Britain, the undivided Indian army had shrunk to a numerical strength of about 500,000. The total complement of the armed forces included a few squadrons of aircraft and a small number of shore-patrol and minesweeping naval units. The soldiers and weapons were divided between the two new states of India and Pakistan in an approximate ratio of two-to-one. In its share, India received the following:[5]

Army	*Navy*	*Air Force*
12 armored corp regiments	4 sloops	2 fighter squadrons
18 artillery regiments	2 frigates	1 transport squadron
76 infantry battalions	16 minesweepers	
	1 corvette	
	4 trawlers	
	4 harbor defense boats	

Reduced to a size of about 300,000 combatants, the new Indian armed forces reflected the shortcomings inevitable in a country emerging from colonial status. The Indian elements of the country's preindependence armed forces were largely the soldiery and noncommissioned ranks. Thus, before the division between India and Pakistan, the Indian officers corps at the level of lieutenant colonel and above was as follows:[6]

Officer Rank	Indian Origin
Major Generals (and above)	None
Brigadiers	4
Colonels	23
Lt. Colonels	240

Similarly, not a single ship nor air force unit had been commanded by an Indian officer during the war. Divided between the two new states, the Indian officer corps was comprised of less than 200 individuals with a potential for military staff level function—but no training, as this was precluded by British colonial policy. The matériel back-up for the military was equally modest, consisting of World War I bolt-action Lee-Enfield rifles, leftover World War II artillery, and stores and spare parts that were cheaper for the allies to leave behind in the subcontinent than to carry back to their home territories. The country had acquired about a dozen ordnance factories but the only lethal weapons produced (when imported gun metal was available) were the Lee-Enfield rifles, light machine guns, and rudimentary artillery pieces. There were no aircraft or naval ship-building facilities other than at the level of repair and refitting, no research or design facilities for armament manufacture, and no national experience in the role of modern armed forces as instruments of state policy.

Complementing the lack in skills, training, matériel, and the relatively modest strength of the armed forces was the attitude of the new Indian political leadership toward the military establishment. There was an element of estrangement between the two since, in nationalist perceptions, the military in the past had discharged the functions of an "army of occupation" in India. While no purges were undertaken, Nehru's government took immediate steps to formally and effectively reduce the preindependence eminence of the armed forces in public life.[7] The military's drain on the public treasury was drastically reduced, its senior officers' ranks in state warrants of precedence (official government hierarchical ranking) were lowered, and its subordinate role relative to civilian-political authority was abundantly clarified. The military organization was to be maintained as an arm of the state in a world of nation-states, but its strength was limited principally to an ability to resist Pakistan. A defense committee of the Indian Cabinet took over the direction of defense policies, but—with the experience of the 1948 Indo-Pakistani war over Kashmir—defense planning provided for little more than Indian army movements across the Punjab plain in case of future wars with Pakistan. With her initial two-to-one superiority in men and arms, India had taken control of two-thirds of Kashmir in 1948. It was assumed that Pakistan would remain inferior to India in military strength and be unable to force her hand in the dispute over Kashmir.

An event of importance during the early years that had a bearing on Indian security occurred in 1950 with the Chinese takeover of Tibet. The high-level North and North-Eastern Border Defense Committee was established in 1951, and submitted its report in early 1953. The effect of its recommendations was mainly to permit a shoring up of border defenses on the Indo-Tibetan boundary. As one writer has noted: "While fully aware of the strategic implications of China's occupation of Tibet, the Indian government . . . responded to the altered Himalayan situation in a manner that must be described as politically discreet, diplomatically cautious, economical of financial and material resources, and projected over a long time."[8] At least up to 1957, the Indian government seemed to preclude a "vigorous and publicized program of Himalayan security measures . . . [fearing that such activities would] compromise the government's professions of friendship and goodwill toward China and provoke the very response which Indian diplomacy sought to prevent—an overt challenge along the long Himalayan frontier."[9]

In a broader geopolitical context, Indian leaders at the time harbored no deep-seated anxieties about the global intent of the United States or the Soviet Union in relation to their country. To obviate any doubts that may have existed, Nehru announced a policy of opening détente with the Soviet Union parallel to the détente that already existed with the Western states. While stating that "we intend cooperating with the United States of America and we intend cooperating fully with the Soviet Union," the Indian prime minister also explained in Parliament that "in accepting economic help, or in getting political help, it is not a wise policy to put all your eggs in one basket."[10] Indeed, the broader justification of what was later termed "Indian neutralism" had been enunciated by Nehru as early as 1931:

> It may be that some will covet her, but the master desire will be to prevent any other nation from possessing India. No country will tolerate the idea of another acquiring the commanding position which England occupied for so long. If any power was covetous enough to make the attempt, all the others would combine to trounce the intruder. This mutual rivalry would itself be the surest guarantee against an attack on India.[11]

From 1947 until today, the broad framework of Indian security policymaking has not diverged significantly from the preceding geostrategic assessment. Somewhat inadequately described now as "nonalignment," India's external policies in the early years following independence searched for ways to advantageously manipulate the country's role and position between the competing socialist and nonsocialist sectors of the international system. While avoiding any formal commitments, beneficial channels of communication were opened with both the latter groups of states to allow for the transfer of techniques, funds, and resources in support of India's economic development. Most importantly, Indian policies attempted to sanitize the level and intensity of interaction with either group according to Indian preferences. Since the Indian ability to pursue its objectives was limited by the power and ability of other states, this reality required India to respond to others' activities as often as to initiate action.

The signing in 1954 of a mutual security treaty between the United States and Pakistan was an event that India could neither prevent nor ignore. Viewing the pact as one that brought the cold war to its doorstep, India first tried to dissuade the cosignatories from proceeding with their plan. When that attempt failed and large quantities of U.S. arms-without-payment started flowing into Pakistan, India responded with purchases of her own from sundry non-U.S. sources (but not at that stage from the Soviet Union). Nevertheless, India lost its earlier two-to-one superiority over Pakistan in armor and aircraft and retained it only in numbers of infantry. For a short period there was some alarm about the possibility of "Pakistani Patton tanks clanking down the Grand Trunk Road to Delhi." However, Pakistan equalled but never acquired a substantial edge over India in attack-weapons systems either on the ground or in the air. As such, Pakistan managed to balance India in the early stages of its security relationship with the United States but never came to possess—probably according to U.S. design—a significant offensive capacity against India. Indian security managers, however, could not assess the U.S.-Pakistani military relationship benignly, and sought to counter the Pakistanis by instituting long-term countermeasures. Under the direction of V. K. Krishna Menon, the foundations for a domestic arms production industry were laid, so that over time Pakistan could be militarily neutralized without requiring heavy changes in future Indian policies to suit foreign weapons suppliers. It was not to become apparent until the end of the sixties that U.S. arms transfers to Pakistan, though eventually amounting to $2 billion and large in absolute terms, would be delivered in annual increments that India could match with its financial resources.[12]

There was, however, to be no period of respite for Indian security policymakers. India's border dispute with China surfaced in 1958, a mere four years after the military alliance between Pakistan and the United States was formalized. By 1962, India had engaged in a disastrous border war with its northern neighbor. In the immediate aftermath of the war, India had appeared to abandon selectivity in the sources of its weapons supplies. Equipment came, in the short term, from sources as contradictory as the United States, the Soviet Union, Britain, West Germany, Canada, and Yugoslavia. But the trauma of the reverses at Chinese hands was to have a long-lasting impact on Indian military planning and preparedness in the future. China, in fact, charged that Western imperialists had allied with Soviet revisionists in support of the reactionary

Indian expansionists. It has since become known that Soviet reluctance to accept fully China's position in the border dispute contributed significantly to the Sino-Soviet split. While that event provided little immediate comfort to India in its military debacle, the fact was of some importance in international terms. It marked the first occasion when the largest communist state in the world openly disagreed with the second largest communist state. Indeed, unconfirmed reports suggest that the Soviet Union withheld the supply of petroleum to China during the hostilities, and soon thereafter, offered to provide large-scale arms aid to India. The offer of Soviet weapons was to be taken up much later, after it had become apparent that the Western powers would supply arms to India only on condition that it compromised over Kashmir with Pakistan and modified its nonalignment to suit Western global objectives.

On the Indian side, long-term defense needs were not assessed on the basis of the ability to purchase arms from abroad but on the ability of local production to meet those needs. This was another issue on which the Western powers demurred at Indian goals while the Soviets supported them. While India went ahead with interim purchases of weapons from wherever they could be acquired, the major change was with respect to a determined effort at creating a domestic arms production capacity. The earlier leisurely pace of developing an armament industry was speeded up and its requirements formalized as high priority items within the general context of planned development. Beginning in 1964, five-year defense plans became rolling adjuncts of the country's five-year economic development plans.

Separate from the long-term projection of its rearmament drive, the Indian government proceeded in the aftermath of the border war to upgrade considerably the numbers, weapons, training, firepower and mobility of the country's armed forces. Simultaneously, a vast program was undertaken to develop a joint civilian-military infrastructure to support the armed forces in any future hostilities on the northern, western, and eastern borders. These activities involved the construction of a large network of border roads and defense lines in the

Himalayan mountains, the training for special mountain warfare of border-security and guerrilla troops, the integration of the civilian and military transport systems, and the establishment of modern defense communication and early-warning radar facilities.

In summary, what had until then been described, even by ministers of the Indian cabinet, as "our parade ground armed forces" were converted into a modern and integrated military establishment. The mission and task of the new armed forces, and hence their functional complement of men and matériel, were defined in precise terms. These were: to provide the country with the ability to simultaneously fight—and dominate—all of Pakistan's military forces and all Chinese forces in Tibet. Any misgivings that may have existed about the new burdens being assumed were soon dispelled as China carried out its first nuclear test in October of 1964. It ensured both that India's conventional rearmament plans would be diligently pursued, and that its nuclear—and, later, space research—activities would be steadily expanded to achieve the technical capacities being acquired by China.

In retrospect, the large-scale increases and modernization of the Indian armed forces following the border war with China probably led to India's next security crisis. Pakistan, aware of the fact that India's armed strength would soon become overwhelming, sought a resolution of the Kashmir problem before time ran out. A twenty-two day war broke out between the two countries in September 1965, within a year of the commencement of India's first five-year defense plan.[13] However, Pakistan was unable to wrest Kashmir from India. The tactical gains by either side during the military campaign were minor, and they were traded off in the peace negotiations that followed at Tashkent.

The Soviet venue and sponsorship of the peace negotiations were themselves unusual. For separate reasons, both Pakistan and India rejected Western initiatives for settling the peace. The Pakistanis were resentful because the United States and Britain withheld military supplies to the two combatants once the war started, thereby hurting Pakistan's U.S.-created military capability more than India's diversified weapons base.

The Indians, on the other hand, had always held the Western states more partisan on behalf of Pakistan. Likewise, the Soviet offer to sponsor negotiations was accepted by the two subcontinental states for contradictory reasons. The Indian case concerning Kashmir had consistently found more favor in Moscow than Washington, and so had their rearmament plans following the border war with China. Pakistan, meanwhile, sought to diversify the sources of its military and political support and complicate those of India by inviting Soviet aid for itself.

The Indians had feared collusive military action between China and Pakistan ever since the 1962 border war. During the campaign in 1965, the Chinese did signal hostile intent, accusing Indian troops of having transgressed into Tibetan territory, and undertaking military maneuvers of their own close to the northern Indian border. While an attack did not materialize, the obvious attempt at joint Sino-Pakistani pressure on India reinforced Indian images of a possible two-front war contingency in the future.

In relation to security assessment and planning, the effect of the 1965 war was to make India that much more determined to implement its extensive rearmament goals. Pakistan's ability to seek accommodation and receive benefits in varying amounts from such ideologically diverse states as the United States, China, and the Soviet Union impressed Indian security analysts immensely. It vindicated the earlier policy assessment that Indian recompense would be best sought by a domestic, and therefore independent, arms industry and military organization.

By the time of the 1971 Indo-Pakistani War, the Indian armed forces had been thoroughly reorganized.[14] The first five-year defense plan, 1964-1969, had been implemented, and the second one had commenced. When the war came in late 1971, India embarked upon the campaign with a superiority over Pakistan in effective-equivalent combat terms of 8-to-1 in aircraft, 4-to-1 in troops, 2-to-1 in armor, and 5-to-1 in naval vessels. More important, all the small arms and ammunition and significant proportions of the aircraft, artillery, armor, and naval vessels on the Indian side were of local manufacture. Staff planning for the war by India was of subcontinental dimensions. Three hundred thousand troops remained on combat alert (and were, if necessary, deployable) on the northern borders, and 500,000 soldiers were used evenly on the western and eastern fronts against Pakistan. The ground movements were integrated with naval attack squadrons, deployed across the Bay of Bengal and the Arabian Sea, and with an air campaign that implemented an average of 500 sorties per day.

Although not recognized by external observers, the war's results were a foregone conclusion to the Indians according to their prior assessments. However, certain events during and before the outbreak could not be predicted, and it is possible that they will eventually carry wider implications than the immediate consequences of the war. The first of these unpredictable events was the initial wavering of Soviet support for India in the months preceding the signing of their Treaty of Peace and Friendship in August 1971. Then occurred the dramatic U.S. opening to China and the secret Kissinger trip to Peking via Islamabad, made with Pakistani help and knowledge. Then followed the statement of U.S. intent, which was delivered to Indian leaders by Kissinger soon after his Peking visit: if India intervened in the East Pakistani situation and China responded militarily along with Pakistan, India could not count on U.S. support.[15] Indian leaders viewed the Kissinger statement as a direct threat carried on behalf of Peking and a visible sanction on behalf of the United States. Finally, during the hostilities themselves, the United States ordered Task Force 74, led by the nuclear-powered and armed aircraft carrier *Enterprise*, into the Bay of Bengal. By all informal accounts, the Indian decision to proceed with a nuclear test was made in the context of the preceding events. Likewise, India's rocket and electronics development programs were sharply upgraded in the months that followed the 1971 war.

The Political Economy of Development and Defense in India

Within a year of independence, the Indian government laid down the broad framework of industrial activity in the new state. As described in the Industrial Policy Statement of 1948, eighteen crucial sectors of industry, including defense, were reserved exclusively for the state-owned public sector. Another forty-six areas of industrial endeavor were earmarked for competitive enterprise by both the public and private sectors of the economy. Everything else, along with agriculture, was allotted to private initiative. Over the past thirty years, only minor changes have been permitted in the mixed economy prescribed for India in 1948, so that today approximately 40 percent of all industrial assets in the country, including core sectors of the national economy, is state owned or state directed. Unlike the practice in countries such as Britain, there is not a single instance of state-owned enterprises reverting to private ownership in India. Indeed, whatever changes that have taken place since 1948 have led only to greater state ownership of industrial production in the country.

In translating industrial policy into practice, successive Indian regimes have been guided by certain premises possessing a law-like sanctity of application. A primal value has been placed on achieving a large measure of self-sufficiency for the national economy. Ideological differences have never been allowed to interfere with trade or aid relationships with other countries. In fact, India's choice in favor of a mixed economic framework has enabled it to draft support for its economic development from both the socialist and the nonsocialist sectors of the international community. The result has been that the socialist states generally have helped in the establishment of India's state-owned enterprises, while nonsocialist states have preferred aiding the nation's private industrial activities. Again, it has been a sine qua non of all Indian contracts for the transfer of plants and techniques from abroad that they provide for the manufacture and not the mere assembly of goods within the country according to a time-bound schedule. An additional requirement has been that Indian personnel be inducted for training in the construction and thereafter progressively in the management of plants set up (or owned) in India by foreign contractors. Finally, there has been an almost religious fervor in encouraging the early creation of a heavy industrial and technologically sophisticated base for India. Abstract economic, political, or ideological arguments that would deny to India the speedy construction of a society that was self-reliant, strong, technologically mature, and the equal of the major powers of the international system have been consistently rejected.

The application of this attitude to defense purposes has meant that the absolute size, diversity, and structural transformations effected in India's modern industrial sector since independence have more significance than their attributes in per capita terms. In that respect, the nominal assets valuation of the industrial economy ranks India tenth in the world, but the volume of industrial production places it in eighth position internationally.[16] Accordingly, India's defense production complex is now the second largest sector of the industrial economy. Its turnover for 1979 was estimated at $1.33 billion in nominal exchange parities, but amounted to two-and-a-half times that figure if valued in terms of the internal purchasing power of the rupee.[17] Direct employment in the defense sector currently accounts for 280,000 workers, and it is assessed that another 1.5 million are indirectly engaged in defense-related enterprises. The production complex comprises thirty-two ordnance factories under the Department of Defense Production, eighteen independent undertakings in the public sector, and fifty other units producing nonlethal items in the private sector.[18]

Despite the importance and size of the defense sector, its development has been undertaken as a normal and not a special part of the industrial economy. Thus, the defense factories have been designed so as to allow them a civilian function as well. At present, 10 to 15 percent of the output from defense factories is for civilian use, and that allocation can be increased during periods of slack demand from the defense services. It is noteworthy that the units remain profitable despite the flexibility of the market, the sharply varying levels of war and peacetime demand, and the multiplicity of end products that have to be turned out. The economic viability of India's defense factories may be partly due to the fact that, since the basic units are wholly owned by the state and the armed forces are thus not placed in the role of customers to powerful private corporations, conversions from war to civilian-related production can be carried out quickly and selectively as the situation demands.[19] Defense inventories have been computer controlled for some years now, so that reasonable forecasts of fluctuations are available to factory managers. This has permitted switches in production from war to civilian-related items whenever needed.

To ensure the future economic viability of its defense industries and for reasons of political affinity, India is likely to enter the Third World arms export market in a modest way in the coming years. A number of Afro-Asian states have asked to purchase Indian arms, and the Indian government appears ready to respond once policy is clarified as to what types of weapons and which countries should fall within the purview of such export projections. Token arms sales abroad, amounting to about $20-25 million annually—mainly to test the market and gain experience—have been undertaken in the past few years. Defense factories have been directed to reserve 20 to 25 percent of their capacities to cater to export demand and diversification.[20] This would mean that India could be selling arms worth $300-400 million at some stage in the future. As indicated elsewhere, the international market value of the weapons sales would be two-and-a-half times their nominal value, once corrections are made for the parity of the Indian rupee. In other words, the export value of 25 percent of India's defense production would amount to a billion dollars annually.

Because of the manner in which the defense production sector has been dovetailed with the nation's industrial economy, its growth and shortcomings parallel those of industry. The main difference between the two sectors is that security assessments have demanded a speedier implementation of the defense-oriented sector since the border war with China than they did before it. While India's gross economic growth rates have averaged between 3 and 4 percent per annum since the fifties, they hide the fact that the industrial growth rate has been about 7 to 8 percent annually over the same period. In the crucial area of machine building and heavy industry, a sector of direct relevance to defense production, the compounded growth rates have been in the range of 10 to 11 percent annually. These figures are fairly similar to China's in terms of the change in percentage, although the absolute size of the Chinese economy was initially larger than the Indian and remains so. Like China, India now produces practically all of its ongoing requirements for heavy industrial plants, machinery, and equipment. Rhetoric aside, China's "self-reliance" appears to be no greater than that of India. In the area of defense production, the comparative position of the two countries is reflected in Table 4.1.

Complementing the structural transformation of India's industrial economy is the substantial reservoir of skilled and technical personnel across a wide range of scientific and applied disciplines. Numbering close to 2 million, and doubling its base every ten years, its numbers are exceeded only by those available in the United States and the Soviet Union. Contradictory as it seems, China, Britain, France, Germany, and Japan have fewer technically skilled people today than does India. It may well be, as an OECD (Organization for Economic Cooperation and Development) study concludes, that India has, to some extent, developed "an autonomous scientific community."[21]

To stabilize the process of scientific and technological change, the Indian government organized in 1971 a separate Department of Science and Technology and a policymaking National Committee on Science and Technology. A Science and Technology Plan now being implemented and integrated with the five-year economic and defense plans has the clear objective of promoting "science and technology and their application to the development and security of the nation."[22] The areas of special endeavor are: energy, mining, metallurgy, heavy and design engineering, chemicals, atomic energy, outer space, electronics, agriculture, housing, health, and education. Of an outlay of $2 billion in nominal exchange parities over the 1974-1979 period—but two-and-a-half times higher if the rupee's internal purchasing power is taken into account—approximately a third has been allocated to atomic energy and space research, and another significant proportion to electronics and computer-related research. As will be obvious, the

Table 4.1. Ability to Manufacture Weapons:
China and India, 1979 (Selected Items)

Item	China	India	Remarks
Space and nuclear			
Nuclear weapons	+	?	
Missiles	+		India has SRBM & MRBM capacity; and IRBM capability in inception stage.
Air			
Aircraft:			
Bombers	+		India has decided not to produce bombers. China is not producing them anymore.
Supersonic fighter bombers	+	+	
Subsonic fighter bombers	+	+	
Transport	+	+	
Trainer	+	+	
Helicopter	+	+	
Aircraft missiles	+	+	
Ground			
Tanks/armored vehicles	+	+	
Guns/artillery/field pieces (radar-controlled and others)	+	+	
Antitank missiles	+	+	
Infantry weapons (rifles, mortars, machine guns, etc.)	+	+	
Heavy trucks	+	+	
Jeeps and patrol wagons	+	+	
Ground support equipment (all varieties)	+	+	
Sea/under sea			
Submarines	+	+	Indian production to begin soon.
Destroyers	+	+	
Missile-carrying boats	+	+	
Fast patrol boats	+	+	
Miscellaneous			
Defense radar	+	+	
Electronics for defense (man-pack transmitters, tracking equipment, etc.)	+	+	

India's Military Power and Policy

allocations to nuclear, space, and electronics research serve the needs of both development and defense.

Distinct from the general scientific and technological effort that is geared to socio-economic development objectives (and defense sophistication in a wider sense), there exists in India a special organization for conventional weapons research, design, and prototype production. Called the Defense Research and Development Organization (DRDO), it manages thirty-four major establishments and laboratories, employs a scientific and technical staff of approximately 6,000, and is concerned with narrowly focused weapons projects. It engages in the duplication, modification, improvement, standardization, and import substitution of defense items. It has worked to develop prototype weapons to suit the local terrain and to create facilities for the local repair and refitting of foreign weapons. Finally, it has encouraged local entrepreneurs to produce spare parts for defense and has undertaken research in special materials and alloys.[23] The actual weapons-development projects in hand remain classified but the general areas in which the DRDO's own seven-year research plan (1972-1979) lays emphasis have been indicated: missile technology, aeronautics, radar technology, underwater naval weaponry, infrared devices, low-level night photography equipment, solid and liquid propellants, inertial navigation systems, sonobuoys, and armored vehicles. The budgetary allocations for the DRDO are probably in ranges of $50-100 million dollars in nominal exchange parities (two-and-a-half times higher with the rupee's internal purchasing power).[24]

While the 1962 border war with China is a natural watershed for Indian threat perceptions, a few crucial weapons projects had actually been initiated prior to the hostilities. As one observer has stated:

> India had made substantial progress in developing local sources of defense equipment, particularly after 1959, on a foundation that was built in less spectacular fashion between 1947 and 1958. HAL's (Hindustan Aircraft Limited) aero-engine division had achieved the distinction of being the first organization in non-communist Asia to manufacture a gas turbine aero-engine. The HJT-16 was the first jet aircraft designed by an Afro-Asian

country without help from either of the two power blocs. The HF-24 project gave India the distinction of being one of only four or five countries to proceed with the development of a supersonic fighter aircraft.[25]

The country already had been manufacturing 80 percent of all the small arms and light equipment for the army, and steps had been taken to establish production lines for heavy trucks, jeeps, and patrol wagons. In 1962, a contract was signed with the British firm of Vickers-Armstrong for the progressive manufacture of a version of the Chieftain tank modified to Indian conditions. In the naval field, the government had acquired two private shipyards for expansion and modernization with a plan to commence the construction of small defense vessels, minesweepers, frigates, and eventually destroyers and submarines.

In large part, the defense production effort prior to the border war with China was in the nature of a learning experience. The means adopted for production were by way of licensed manufacture of prototype equipment. The credit for India's being able to arrange such agreements in the late fifties—a time when they were extremely difficult to obtain—is due solely to V. K. Krishna Menon, who had taken over as defense minister in 1958. Under his stewardship, licenses were acquired for the production of Gnat interceptor aircraft (U.K.), Alouette helicopters (France), L-70 antiaircraft guns (Sweden), Vijayanta tanks (U.K.), Shaktiman trucks (West Germany), and 106 mm recoilless guns (United States). Menon also sought to persuade the British to permit the manufacture of their Lightning fighter aircraft in India. When the request was rejected, India turned to the Soviet Union, and in August of 1962 reached agreement for the manufacture of MiG-21 aircraft within the country.

It seems clear that, from a very early stage, the Indian defense establishment was concerned with the import of technological know-how as a condition for the purchase of defense equipment. The initial approaches were usually made to Western countries, and only upon their refusal were the options with the Soviet bloc countries used. In both instances, the goal of domestic manufacture was diligently pursued. This policy frequently led to delays in the availability of foreign weapons to India as its negotiators engaged in lengthy discussions and dangled the carrot of weapons purchases before prospective licensors in return for the delivery of technology. India's bargaining power in such matters was obviously limited in the context of the times, but to the extent available it was employed quite deliberately for long-run gains.

The security policymaking exercise in India has always been confined to small groups of people within the government. Only in recent years has a modest tradition of public debate developed within the country on matters of defense. In the fifties, the subject did not excite the imagination of even senior ministers to the government. Indeed, the Ministry of Defense was considered a sinecure appointment by those upwardly mobile in the national political hierarchy. As a result, financial allocations for defense generally, and weapons production specifically, took their low turn in central budget disbursements. There was an abrupt turnaround after the 1962 experience with China, when defense expenditures and procurement became unquestionable elements in all succeeding budgetary- and economic-plan investments.

Given the current size and weapons inventories of India's armed forces, it is instructive to see the manner in which the force modernization process was undertaken. First, it was decided to increase military capabilities through national effort over the long term rather than by resort to panicky purchases of foreign weapons. Second, the new force structure was to be determined according to Indian objectives and by the local staff planners. Advice was invited from friendly states, but their offers to provide India with large numbers of military advisors and staff planners were rejected. Third, the inflationary impact of escalating military expenditures was to be diffused by monitoring their release according to five-year defense plans. The latter, in turn, were integrated with the ongoing five-year economic development plans. Fourth, foreign exchange expenditures were to be kept to a minimum. Fifth, all weapons acquired from foreign sources were to be paid for. Sixth, the modernization program was to proceed in a series of steps, with emphasis on expansion of the army and air force in the initial stage, improvements in the technical and managerial organization of defense in the next stage, and increases in the firepower and mobility of all three services in the last stage. Finally, a rolling-plan concept was applied to the defense build-up to allow for a continuous review of the programs under implementation.

India's first five-year defense plan, 1964-1969, led to a doubling of the army to 825,000 soldiers, and to equipping it with modern weapons. The structure of the land forces was recast so that its numbers and weapons could be rapidly expanded beyond authorized strength at short notice. Likewise, the strength of the air force was doubled to forty-five squadrons of front-line jet aircraft, and air defense and communication facilities were modernized. The other specific objectives of the first defense plan were: replacement of over-age ships of the navy; extension of the border roads network; strengthening of the indigenous defense production base; and improvement of the defense organization structure. The expenditure on the implementation of the plan was placed at 50 billion rupees ($10 billion at pre-1966 rupee devaluation rate), with a foreign exchange component of 7 billion rupees (or a little over $1 billion).

The guidelines in planning for defense production focused on the need for efficient and modern weapons. The targets for self-sufficiency were delineated according to time-bound schedules for different types of weapons systems. Items of slow obsolescence were to be stockpiled, and research and development would emphasize defense equipment suited to Indian conditions. Capacity in the civilian sector was to be utilized to the maximum extent possible for the provision of nonlethal stores. High standards of managerial and technical skills were made mandatory to the running of defense factories.

The overall direction of the new defense effort was given over to the inner Defense Committee of the Indian Cabinet. Executive responsibilities were shared functionally between the National Defense Council and the specialized Defense Research and Development Council. By 1965, a high-level Defense Planning Cell was created to deal with all wider aspects of planning that have a bearing on defense in both the medium- and long-term aspects, and to maintain constant liaison with the (Economic) Planning Commission. In 1966, the Department of Defense Supplies was reorganized to undertake the location of domestically produced substitutes for all imported components of weapons, to encourage individual firms to supplement the government effort at import replacement, and to coordinate private sector scientific and technical research with the work of the Defense Research and Development Organization.

The second and third defense plans (1969-1974 and 1974-1979) called for approximate expenditures of 75 billion rupees each (or $10 billion for each of the plans at the post-1966 rupee devaluation exchange rate). Thus, by the end of the current decade India's rearmament and force modernization effort, begun in 1964, will have cost some $30 billion, or $2 billion per year over a fifteen-year period. The financial outlays for the next defense plan, now underway, will at least equal but probably exceed the expenditure rates of the previous years. Its main objective, according to statements by defense officials, is to phase out by 1985 all existing Indian dependence on foreign licensors for the supply of crucial components of weapons systems otherwise manufactured in the country. A second aim is to invite greater support of the private sector industries in the national defense effort by encouraging them to set up ancillary, spare parts, electronic, and specialized chemical manufacturing units. The decentralization of defense production, with a share allotted to the private sector, has already made substantial headway. More than 50 percent of the annual value of Indian military production for nonlethal items now originates from private firms. The changes in policy are, however, calculated, and still disallow private firms from entering into the manufacture of lethal weapons systems. The third area of emphasis in the new defense plan relates to achieving competency in the Indian design and development of complex weapons systems.

It is likely, under the circumstances, that the annual value of Indian defense production will rise significantly beyond the current level as production lines are established for the high capital-cost weapons now in various stages of inception. These include: a domestically designed main battle tank for the army by 1980; an Indian-designed, missile-firing, and antisubmarine destroyer for the navy by 1982; and completion of the prototype of HF-73, a locally designed, short-take-off-and-landing, multirole combat aircraft with twice the speed of sound for the air force by 1985. Other important projects relate to the development of hydrofoil and hovercraft vessels, large-sized helicopters, electro-optical equipment, miniaturized computers, heavy marine-diesel engines, super alloy compounds and special steels, and a variety of missiles for ground, naval, and air combat. A noteworthy event last year was the Indian development of a supersonic, remotely piloted vehicle. Although currently used as a target drone by the air force and the navy, its technology is similar to that employed in cruise missiles.[26] Special naval-nuclear teams are also known to be engaged in research on the development of nuclear propulsion systems for undersea craft. It is probable, moreover, that India will soon manufacture long-range, missile-deploying corvettes of a type similar to the Nanuchka-class boats, which it recently acquired from the Soviet Union. These vessels are mounted with a variable mix of ship-to-ship missiles and have a range of 1,600 miles.[27] Finally, the chairman of the Indian Space Research Organization stated last year that the country possessed the capability to build intermediate-range ballistic missiles (these have a range of 1,500-4,000 miles). He further suggested that "we would have had larger launchers if the military were giving the push (to the space program)."[28]

Indian planning for defense production does not bar the country from buying into proven technology from elsewhere, as long as the contractual terms fall within the policy guidelines mentioned earlier in this study. In this respect, the country now appears to have moved from the position of a supplicant to that of a courted customer. In the market recently for the licensed production of a deep penetration strike aircraft, India had the choice, among Western suppliers, of the Swedish Viggen, the French Mirage F-1, and the Anglo-French Jaguar. The terms offered by the vendors of all three aircraft were extremely favorable from the Indian viewpoint. It has been learned, meanwhile, that the Soviet Union made a counteroffer to India of a choice among the MiG-23, the Su-20, and the Su-22 aircraft. The selection, eventually, of the Jaguar in a $2 billion deal was based on a number of considerations, the most important among them relating to the scope and speed of the transfer of the aircraft's technology. Additionally, the agreement included the production in India of all the spare parts for the Jaguar. In this respect, Indian experience with the earlier MiG-21 arrangement had been mixed, and that factor contributed to the decision to diversify Indian sources of defense technology. The change, however, does not spell severance of the Soviet military connection, which continues to remain strong. Agreement was simultaneously reached with the Soviet Union for the transfer of blueprints and technology to manufacture the more versatile MiG-BIS plane in India as a follow-on to the standard MiG-21 aircraft already built in India. It is interesting to note, moreover, that the Jaguar deal, apart from the political advantage of spreading expenditure between Britain and France, envisages the transfer to India of the same Turbomecca aero-engine that is to power the Mirage 2000, the next generation of French fighter aircraft. The British, meanwhile, are also providing a small number of Harrier jump-jets to India as part of the total package, and its technology will aid in the development of the indigenous HF-73 short-take-off-and-landing aircraft.[29]

India is currently evaluating offers from five different countries for the licensed production of submarines within the country. Its choice among prospective licensors will depend on considerations similar to those attending the Jaguar aircraft agreement. Indicative of the country's improved bargaining position in such matters, the Soviets are reported to have expressed a willingness to convert India's existing submarine fleet to nuclear propulsion.[30] Local interest, however, is in conventionally powered, medium-sized vessels, and the plan is to import a complete submarine and then to build the remainder under license. In political terms, it is clear that India continues to benefit from its amicable relations with both the socialist and nonsocialist segments of the international system. Indeed, the only significant omission from the country's military production relates to the United States, which, for reasons of its own, demurs from participation in Indian defense ventures. There is no reason to believe, however, that U.S. policy is immutable. If it changes, the Indians would most likely be in the market for purchase of U.S. defense technology.

In overall terms, then, India will have made a capital investment of some $40 billion over a twenty-year period ending in 1985, to acquire a relatively independent and sophisticated defense production base and a modern military establishment. Approximately 11 percent of the government's research and development expenditures (or, as noted, $100 million) are now being devoted to the development of weapons systems. It is estimated that a further capital investment of $1 billion will put the Indian armament industry on a stable basis for local and export purposes. Spread over a number of years, the annual allocations would be small and well within the country's means to apportion without undue pressure on alternative funding needs.

Separate from the capital investments in defense are the expenditures incurred by India for the current maintenance of its armed forces. On an average, 16 to 20 percent of the Indian federal budgets (that is, excluding state budgets) have been devoted to the annual upkeep of the three services. The ratios of the defense burden as a percentage of GNP are indicated in Table 4.2.[31]

Expenditure patterns have fluctuated between 3 and 4 percent of GNP from 1976 to 1979, and on a per capita basis they amount to $4 annually. On a comparative scale, Pakistan's defense expenditures as a percentage of GNP have varied between 7 and 10 percent, and expenditure per capita at $9-11. While exact data for China are not available, Western assessments are that the Chinese spend about 10 percent of their GNP on defense at a per capita burden of $15.[32] Soviet estimates are much higher. Thus, in relation to their supposed adversaries, the Indians have devoted appreciably less funds to defense purposes. Indeed, on a worldwide comparative scale, Indian defense expenditures figure in the lowest quartile of national defense allocations.[33] In theoretical terms, therefore, India has the facility to increase its defense expenditures over a

Table 4.2. India: Defense Expenditures
as a Percentage of GNP, 1950-1978

Year	Defense Budget ($ million) 1970 price & exchange	Percentage of GNP
1950	334.4	1.6
1951	370.4	1.8
1952	551.1	1.7
1953	548.2	1.7
1954	585.6	1.8
1955	610.2	1.7
1956	607.1	1.7
1957	730.4	2.1
1958	723.2	2.0
1959	674.2	1.9
1960	677.6	1.9
1961	728.0	1.9
1962	1003.7	2.6
1963	1642.5	3.8
1964	1607.6	3.6
1965	1567.6	3.6
1966	1480.1	3.4
1967	1373.2	3.0
1968	1429.0	3.1
1969	1511.9	3.0
1970	1558.2	3.0
1971	1743.9	3.7
1972	1995.5	3.7
1973	1929.0	3.0
1974	2180.0	3.0
1975	2660.0	2.8
1976	2812.0	3.0
1977	3117.0	3.1
1978	3571.0	3.1

wide margin if it wanted to match the percentage burdens of its commonly perceived adversaries or those of the major powers.

It is unlikely, however, that India will increase its defense burden over present rates. The trade-offs in capital or current investments between the needs of defense and development appear to have stabilized at existing levels. Once security requirements were perceived as greater, it was decided to add expenditures over a time-curve rather than to bunch them in a few short years. As a result, the annual burden of increased defense expenses could come from increasing budgets instead of being diverted from existing development projects. Of course, there was some substitution of funds earmarked for development to defense, especially in the initial years of defense modernization, but the negative effects of the change were relatively low and absorbed quite smoothly in total governmental expenditure patterns.[34]

The present long-range programming pattern adopted for internal defense reorganization, particularly with regard to weapons procurement, has an analogy in the policies that governed earlier arms purchases from abroad. Following the war with China, the immediate need for modern weapons in expectation of the contingency of a two-front war had been placed before the United States and Britain. As pointed out elsewhere, their response was conditional to changes in Indian policy over Kashmir, India's relations with the Soviet Union, and India's socialist-oriented public sector biases for the internal economy. A request for submarines—one of which had been supplied to Pakistan—was turned down, and so was the Indian plea for the initial supply and later production of a front-line aircraft within the country. Under these circumstances, India turned to the Soviet Union, which became, in 1964, the country's major foreign supplier of arms, and since 1968, the source for crucial items of licensed manufacture. The Western connection was not broken, as equipment valued at about $70 million was received from the United States for some of India's newly formed mountain warfare divisions and for the construction of a distant early-warning radar system along the northern border. The modest nature of Western support, however, was more than compensated for by the Soviets. The latter not only opened up their defense inventories to India, but agreed to supply weapons against payment in rupees. India could subsequently recoup its currency by selling civilian goods to the Soviet Union. The economic burden of such an arrangement was no less than it would have been through more conventional payment terms. But for a country short of hard currency, bereft of weapons sources, and limited in its options, the reverse-purchase agreement with the Soviet Union for defense supplies was of incalculable advantage. In its most significant aspect, the financial agreement governing arms purchases from the Soviet Union allowed India to continue using its (then) meager foreign exchange resources for the purchase of civilian plants, machinery, and technology from abroad. Thus, the country's economic development, particularly its industrialization drive that required resource transfers from the international economy, continued as before despite the new burden of defense. Additionally, and over the long term to mutual benefit, the deal with the Soviet Union brought into existence a new market for the sale of Indian goods and thereby aided the development of the national economy. Since the sales and purchases cancelled each other on both sides, neither India nor the Soviet Union owe any monies to one another. Meanwhile, the Soviet Union has become India's second largest trading partner, and the overall economic relationship continues to grow.

India's foreign arms purchases from all sources for the period 1965-1975 are estimated at approximately $2 billion. The annual deliveries in monetary terms are indicated in Table 4.3.[35] Thus, India's average annual import bill for arms purchases has been $200 million and the rate of intake remains the same for recent years. For reasons discussed earlier, the allocations have not been unduly burdensome to India and more or less match the value of Pakistani weapons acquisitions. The major difference lies in the fact that India has paid for all its foreign arms whereas Pakistan received on grant bases its $2 billion worth of American arms (until 1966) and subsequently an unknown quantity of weapons from China, while paying for those it acquired from elsewhere.

The Impact of India's Political and Military Choices

The events of 1962 on India's northern border mark the great divide in its attitudes toward defense. Prior to the clash with China, the country possessed a colonial-style and colonial equipped armed force whose requirements, though not neglected, were low in the scale of priorities. No military threats, other than the one from Pakistan, were perceived as endangering India's territoriality. Specifically, there was no

Table 4.3. India: Foreign Arms Purchases, 1965-1975

Year	$ Million (constant)
1965	189
1966	377
1967	133
1968	212
1969	171
1970	114
1971	257
1972	216
1973	180
1974	117
1975 (est.)	100
Total	2066

reason to fear an attack by states beyond the Indian subcontinent. India's refusal to participate in military alliances, while irritating to the Western powers, was not a casus belli to attack India. The country's stubbornness in the matter helped to reduce early Soviet suspicions of Indian intent. It also provided India with a legitimate basis to expect a friendly rather than a hostile response from the Soviet Union. Interaction with the new China was premised in a similar manner. Indeed, a special empathy was manifested toward the Chinese as fellow Asians recently arisen from semicolonial status. China's differing political system did not trouble Indian decision makers, who generally assumed that their northern neighbor would temper its revolutionary zeal with some regard for the common condition of other excolonial Asian states. The Indian approach, if too sentimental, was not an invalid basis for opening relations with China. The alternative lay in reacting to China from the beginning with a hostile stance, as did the Western states. That would inevitably have required India to rearm with Western weapons, and as a consequence to participate in their military alliances. It follows that India would then have incurred the wrath of the Soviet Union, which at the time was China's fraternal ally. In effect, the country would have had to confront the two most powerful states in Asia, one of them a global power, and both of them in its immediate vicinity.

The compulsions of geopolitics are immense in the Indian case. The border war with China only highlighted them. Since 1962, India has responded to the realities of geographical location. China's recent actions in Vietnam, with historical analogies drawn to the border war in 1962 by senior leaders in Peking, serve to remind Indians that China has imperatives of its own for which force will be used. Despite the recent thaw in Sino-Indian relations, there is no indication of flexibility on either side concerning the border issue. As a result, it is almost axiomatic that India will continue to increase its military capability. A tacit dimension of that effort, not openly expressed by policymakers, requires that India achieve some form of military parity with China. The concept does not demand that the country match China in numbers of soldiers, weapons, and aircraft—although that is not impossible—but that its armed strength be sufficient to dominate the Himalayan border. Given China's military needs on its borders with the Soviet Union, and the distances from the country's interior to Tibet, it is relatively within Indian means to achieve conventional-force parity with its northern neighbor.

To some extent, the equivalence with China in the above respect has already been achieved. India now has more mountain warfare troops and firepower in place along the Himalayas than does China. Reports suggest that the official claims about the strength of the Indian army at 1.2 million are considerably understated.[36] Further, the teeth-to-tail ratio of the land forces can be adjusted so as to release substantially more combatants than in normal circumstances. There are 200,000 regular reserve troops. While the official reckoning of India's paramilitary forces is also 200,000, that number excludes 1 million home guards who are also trained and drilled with military weapons and in military formations. Finally, there exist ex-servicemen's units, a national cadet corps, a territorial army, and a national volunteer force, whose total numbers would be in excess of 500,000 and all of whom are deployable with a modicum of training. Thus, in the short term but for an extended land campaign, India has the ability to deploy at least 3 million soldiers. In a lengthy campaign—which is unlikely due to the nature of current weapons and the times—India could resort to conscription, which it does not have at the moment.

India's air force consists today of about 1,000 jet aircraft, of which aproximately 700 are organized in 45 combat squadrons. The number of combat squadrons is being raised to 64, which means that the country will soon have 1,500 front-line aircraft, inclusive of reserves, in service. These figures exclude about 1,000 other nonjet aircraft that are used for support duty by the army and the air force. It is also apparent that India's combat aircraft are more modern than China's. Most of the latter's aircraft consist of MiG-17s and MiG-19s.[37] China's attempt to build a version of the MiG-21, named the F-9 Shenyang, appears to have been unsuccessful, as only a handful are in service with the Chinese air force. Perhaps China intends to use the Spey engine, for which a licensed production contract was signed with the British firm of Rolls Royce, to improve the performance of its F-9 aircraft. However, unless China is soon able to acquire Western technology for its aircraft industry on a par with what India is already in the process of obtaining, China's air force will lag behind India's in terms of new generation aircraft.

Given the sea distances between India and China, their respective navies are not of much consequence in a hypothetical situation of hostilities. China's fifty submarines are an ostensible threat, but since they are diesel-powered they would be open to detection by modern sonar tracking devices. India has this requisite equipment and technology. It has, moreover, recently acquired long-range antisubmarine and reconnaissance aircraft that, in concert with the Nanuchka corvettes of the Indian navy, would make it very difficult for conventional submarines to operate in the seas around the subcontinent. In addition, the Indian navy is presently implementing a separate ten-year development program. That special effort is indicative of the government's intention to compensate for its earlier neglect of naval requirements against the more pressing needs of the army and air force. Thus, while the navy's allocations out of the revenue portion of defense budgets is only 10 percent, it is receiving almost 50 percent of the funds on the capital account.[38] The main objective of the ten-year plan is to provide India with two blue-water navies, one each for the eastern and western Indian Oceans, able to operate independently of one another. The second objective is to augment the shore-based infrastructure to support the navy's deployment beyond coastal waters. Toward that end, Bombay in the west, Cochin in the south, and Vishakhapatnam on the east coast are being developed into large naval bases. India has also been constructing for some time now one of the biggest naval bases in the Indian Ocean. It is situated at Port Blair in the Andaman Islands, which are owned by India and lie athwart the Malacca Straits connecting the South China Sea with the Bay of Bengal. Consideration is being given to the establishment of a second base in the western Indian Ocean in the Indian-owned Laccadive and Minicoy group of islands. Twenty-five submarines are to be added to the existing fleet of eight submarines, with four more on order, and a larger number of locally made, fast missile-deploying boats of the Nanuchka type will augment the coastal forces. Increases are projected in the inventory of shore-based, fixed-wing aircraft of the fleet air arm, and naval reconnaissance and antisubmarine Ilyushin-28 aircraft with an arc-range of 1,000 miles have already been put into service. A new Indian-designed destroyer is currently being constructed. It is the first of a series of modern warships that will be outfitted with antisubmarine warfare equipment, helicopters, a variety of missile batteries, long-range radars, and eventually, a satellite communication system. Guided-missile cruisers also figure in the navy's plans.[39] The enlargement of the Indian navy reflects concern about the expanding deployment of superpower fleets in the seas around the subcontinent, and has little to do with China. But the increases in Indian naval strength appear sufficient to contend with any hypothetical Chinese naval presence in the Indian Ocean.

The obvious imbalance between China and India exists in the former's possession of nuclear strike forces. However, after India's nuclear explosion of 1974, it is generally believed that the country has the ability to make nuclear weapons. India has refused to accept full-scope safeguards—that is, a countrywide monitoring of all its nuclear facilities—despite a variety of pressures and sanctions.[40] While earlier government pronouncements had suggested Indian abstinence from nuclear tests, current reports quote the Indian prime minister as distinguishing between "nuclear blasts for tracing natural resources and explosions for unidentified purposes."[41] The former, apparently, are not barred. The country, moreover, has not neglected research in materials required for thermonuclear experiments, and informal opinion among Indian nuclear scientists is that they could undertake fission-fusion-fission tests if told to do so. Perhaps no branch of Indian industrial and scientific endeavor has been developed in as autarkic a mold as the country's nuclear establishment. While civilian-oriented, the competencies achieved by India in that field could simultaneously support military and peaceful purposes. Its diversified base is certainly superior to China's. For instance, India builds its own commercial nuclear power stations whereas China does not.[42] Similarly, India completed the construction in 1977 of Asia's largest (and first nationally built) variable-energy cyclotron, while China contracted in 1979 to acquire its first similar machine from the United States. Again, India is one of eight countries in the world, and the only one in the Third World, with an ongoing fast breeder-reactor program. Finally, Indian scientists have for some time been engaged in research with lasers, isotope-enrichment techniques, magneto-hydrodynamics, mixed-oxide nuclear fuels, and plasma physics. They have also begun the construction of the first "Tokamak" machine for fusion experiments in the Third World.

In terms of nuclear delivery systems, it is believed that China possesses thirty to forty medium-range missiles with an operational range of 600-700 miles, and an equivalent inventory of intermediate-range launchers with a 1,500-1,750 mile radius of action. It is assumed that, having launched several satellites from a multistage missile of a range of 3,000-3,500 miles, China may have deployed some of those missiles. There is, further, speculation that developmental work continues on missiles with longer ranges.[43] India's rocket capability appears modest in comparison with China's but it is probably not too far behind in technical terms. Indian space research establishments have tested and developed inertial guidance systems, telemetry equipment, rate gyroscopes, heat shields, nose cones, electronic payloads, miniaturized onboard computers, and a variety of high specific-impulse solid and liquid propellants.[44] A locally fabricated 300-kilogram satellite was launched in 1975 under an arrangement with the Soviet Union using a Molniya rocket, and a second larger satellite was launched this year. Also in 1980, India's own satellite launch vehicle, consisting of a four-stage, eighteen-ton rocket designated the SLV-3, will be used to loft a multifunctional satellite into orbit. One of its objectives is to gain experience with long-distance, high-resolution photography. Another Indian space activity open to direct military application is the successful testing of clustered rockets using single boosters. The current phase of India's space program, to be completed by 1985, is designed to enable the lifting of 1,200 kilogram payloads into geostationary earth orbits of 40,000 kilometers in space. Missiles are designated as "intermediate" when they can cover distances of 1,400-4,000 miles. Indian claims to such a capability, quoted earlier, probably refer to missiles with approximate ranges of 3,000 miles.[45]

It is interesting to note that the expenditures on India's nuclear and space programs are amortized against civilian purposes. Commercial nuclear power is profitable and competitive in Indian conditions, and the space activities are to be defrayed against point-to-point communication and satellite instructional television programs. In both instances, therefore, defense options are being provided without a diversion of funds from development objectives.

The exercise in comparing Chinese and Indian defense capabilities does not insinuate the inevitability of a military confrontation between the two. It implies that the Chinese would find it difficult to repeat their military performance of 1962 against India. The latter's tactical military plans for the future defense of the northern borders now include ingress into Tibet as a riposte to any Chinese incursions in Indian territory. The second implication of India's military buildup in relation to Chinese capabilities is that the conventional power balance within the subcontinent has changed unalterably in India's favor. In the joint context of the Himalayan and Pakistani borders, the Indian army can be described functionally as composed of two armies: a mountain army facing north and a plains army facing west. Both the armies are independently superior to the forces they oppose. Staff planning provides for an interchange of soldiers and weapons between the two fronts in a relatively short order, and succeeding defense lines and roads along both frontiers stalemate the military situation and give easier access to resupply. In a limited land war, India has probably acquired the ability to defend itself without outside support. This situation will exist independently of the possibility that Pakistan might acquire external military assistance in significant quantities. It will also obtain vis-à-vis China's capacity, given its total security needs, to apply force against India.

The stabilization of threats across India's land frontiers is not duplicated with respect to the country's maritime boundaries. India's peninsular shape gives it one of the longest national coastlines in the world, and historical memories of the European colonial assault on India from its seaward side run deep. Thus, the unfolding plans of the superpowers for permanent naval deployments in the Indian Ocean have caused major alarm among local decision makers.[46] Analogies are drawn between the nineteenth-century Western onslaught on the region for spices and the twentieth-century Western interest in another local commodity, oil. While India does not possess any resources coveted by other states, its territories skirt the Persian Gulf states that contain the richest trough of oil in the world. The United States has, apparently, decided to create an Indian Ocean command with a special fleet of warships stationed permanently at a large base developed on the island of Diego Garcia.[47] That island lies 800 miles due south of India. The Soviet Union has also sought permanent naval bases in the Indian Ocean, and with the new U.S. plan for Diego Garcia, it is unlikely that the Soviets will avoid for too long an equalizing presence in the region. Thus, an ocean that had remained a backwater for the past thirty years is now in the process of steady militarization. More importantly it is in the process of direct inclusion in the strategic-balances equation of the global powers.

While oil may have been the catalyst in reviving great-power interest in the Indian Ocean, a series of other regional events extend the purview of their military activities from the ocean itself to the lands bordering it. The changes of regime structures in Afghanistan and Iran, the war in Yemen, and the developing conflicts in East Africa are symptomatic of the future turbulence expected in the Indian Ocean region. Indeed, an "arc of crisis" is now identified by great-power analysts as stretching from India's border with Pakistan, through Afghanistan, the Gulf states, the Arabian peninsula, and to the horn of Africa. India has had to contend with a troublesome insurgency of its own in the eastern region bordering Burma, and for years the rebel groups have been trained and supplied by the Chinese.[48] Information received by India indicates that most of northern Burma is overrun by the "white flag" and "red flag" factions of the Burmese Communist party, and effectively beyond the control of its own government. The same might be true of the northern portions of Thailand and Laos. The distance between eastern India and Laos is less than 600 miles, while the intervening territory of thick jungles and north-to-south mountain ridges provides ideal terrain for guerrilla warfare. Given China's new relationship with the United States, Indian strategists expect the latter to accommodate the former's interests and activities across Southeast Asia. Thus, it is not known as to when another "arc of crisis" might erupt to the east of India.

In consequence, Indian policymakers feel that a profound change is taking place in their total strategic environment involving the direct interests, and perhaps the direct interventions, of the United States, the Soviet Union, and China. In that framework of strategic evaluations, Indian concern now exists with the weakness of their subcontinental neighbors relative to their own strength. It is felt that innovative forms of support might, in future contingencies, be pressed upon the other states of South Asia. Besides, the lineup of the three great powers in an actual situation would remain quite unpredictable. With such premonitions, Indian analysts are prone to remember their constricted situation prior to the 1971 war with Pakistan. For several months the Soviet Union wavered in its support, while the United States and China ranged themselves against India.

To ward off a repeat of the intervention attempted in 1971, Indian naval strategists are now integrating their shore-based capability with their capacity on the high seas. The Indian peninsula juts into the Indian Ocean in such a manner that 50 percent of the ocean stretches to the east, west, and south lie within 900 miles of the country's coastline. The configuration of the country, in effect, makes it a huge aircraft carrier positioned permanently in the Indian Ocean. This is a geographical advantage that Indian naval planners are going to exploit in a number of ways. For instance, they are currently in the process of establishing batteries of a new Indian-designed, shore-to-ship missile of undisclosed range along the country's coastline. Again, a long-range radar network is being erected and is designed to detect hostile ships (and also cyclones) far from the country's shores.[49]

A secondary line of naval defense in the initial stages of inception is India's Coast Guard Organization. Under a six-year plan and with an expenditure of $125 million (two-and-a-half times that amount in Indian prices) the embryo of a coastal force is being formed. It will consist, in the early stages, of approximately twenty-five ships of various types, most of them built within the country.[50] The costs of this enterprise are written off against civilian expenditures, since the coast guard's primary functions are to protect India's off-shore oil installations near Bombay, contend with the extensive smuggling operations mounted from the Gulf sheikhdoms into India, and undertake rescue operations. However, it is to be expected that, like other states, India will provide for the integration of the coast guard with the navy in situations of emergency.

Despite the variety of activities undertaken to increase its power, India's conventional military reach remains basically limited to the subcontinent. The need, if it arises, to extend the country's military power beyond South Asia could be met only through a deliberate decision to acquire a nuclear strike and delivery force. No such decision appears to have been taken so far. Items and activities leading to the purchase of a nuclear force are readily recognizable, and Indian effort to that end would be quickly revealed. The question remains moot for the moment and dependent upon a number of uncertainties. Prime among these are: the likelihood of a Pakistani acquisition of nuclear weapons; the presence and activities of the superpowers in the Indian Ocean; the incidence of external interventions in the affairs of the littoral states; increasing wars in regions adjacent to the Indian subcontinent; and negative developments in India's relations with China.

In essence, India's security interests lie within the subcontinent, and the primary objective is to protect the area against outside interference. The country has little reason to exacerbate relations with either of the two superpowers as long as its own ideological preferences and the subcontinent's security are not unduly threatened. The same attitude exists with respect to China, except that after the experience of 1962, India appears reluctant to confront the Chinese across a negotiating table without an equalizing military capability. Indian strategic plans do not, however, override a settlement of the border dispute in which concessions are offered by both sides. Within the subcontinent, India cannot avoid the charge of seeking hegemony, given its size, potential, and now its military capacity relative to that of its neighbors. There may be more rhetoric than substance to that charge. Despite its looming presence in the subcontinent, India could hardly dominate Pakistan with its 75 million people, or Bangladesh with its 80 million citizens.

Conclusion

Indian security managers now view nonalignment with military power as maintaining the benefits that existed earlier for their state without a basis in power. Thus, the substance of India's strategic policies has become somewhat indistinguishable from the policies of other large states. In a broader strategic context, one can surmise that the exercise of nonalignment policies by a state of India's size could be nothing other than a manipulation of other nations' power and resources toward its own power acquisition. Furthermore, there is nothing unusual or unique about the policies adopted by India in the pursuit of its security goals within a framework of nonalignment. China appears to have had similar objectives behind its policies, except that in its case the descriptive term "nonalignment" is less appropriate than the phrase "lean to one side." China is now leaning toward the Western side, but there are few guarantees that its inclination will not change again. What is clear about both the Asian states is that neither has been willing to accept the world view of other large and powerful countries in the determination of its own internal and external strategic environment.

If India is perceived as a future major-power contestant, then its rise to such status will have been relatively painless compared, for instance, to China's. The change will have taken place with the support rather than the extreme hostility of the existing superpowers. That the process of status-change might also be managed with the presence of a relatively open political system could be an unusual event.

Notes

1. *The Military Balance 1978-1979* (London: International Institute for Strategic Studies, 1979). The ranking emphasizes regular army formations and equipment, combat air force units, and inventories of medium and large warships.

2. Comparative data compiled by author. See also, Peter Lock and Herbert Wulf, *Register of Arms Production in Developing Countries* (Hamburg: Study Group on Armaments and Underdevelopment, 1977), section on India, pp. 94-103.

3. For a fuller discussion, see Onkar Marwah, "India's Nuclear and Space Programs: Intent and Policy," *International Security* (Fall 1977):96-121.

4. Bisheshwar Prasad, General Editor, *Official History of the Indian Armed Forces in the Second World War, 1939-45*, 18 vols. (Delhi: Orient Longmans, 1962), combined Inter-Services Historical

Section, India and Pakistan. For an excellent analysis of the growth and role of the Indian army since the nineteenth century, see Stephen P. Cohen, *The Indian Army: Its Contribution to the Development of a Nation* (Berkeley and Los Angeles: University of California Press, 1971).

5. Ibid.
6. Ibid.
7. See, for a succinct analysis, P. R. Chari, "Civil-Military Relations in India," *Armed Forces and Society* 4 (November 1977):3-27.
8. Lorne J. Kavic, *India's Quest for Security* (Berkeley and Los Angeles: University of California Press, 1967), p. 61.
9. Ibid.
10. Publications Division, Ministry of Information and Broadcasting, Government of India, *Independence and After: Speeches of Jawaharlal Nehru, 1946-1949* (Delhi: Publications Division, Ministry of Information and Broadcasting, Government of India, 1949), pp. 205, 217.
11. Quoted in, Kavic, *India's Quest for Security*, p. 23.
12. U.S. official sources admit military assistance and sales to Pakistan from 1950 to 1973 of only $672.5 million and $89.1 million, respectively. See U.S. Department of Defense, Security Assistance Agency, *Foreign Military Sales and Military Assistance Facts* (Washington, D.C.: Arms Control and Disarmament Agency, 1974), pp. 17, 19. Indian estimates discount the above figures, interpreting them as a function of the procedure by which U.S. military grants and sale items are declared as "excess defense equipment," or surplus hardware, and marked down in value for accounting purposes.
13. For a detailed study of the 1965 war between India and Pakistan, see Russell Brines, *The Indo-Pakistani Conflict* (London: Pall Mall Press, 1968).
14. For a review of the manner in which India conducted the military campaign in 1971, see Robert Jackson, *South Asian Crisis* (New York: Praeger, 1975).
15. From interviews with senior Indian officials.
16. A. N. Prabhu, "Engineering Industry Makes Rapid Strides," *Indian and Foreign Review* (February 15, 1979):15-18.
17. See "Government to Decentralise Defence Production," *The Indian Express* (Delhi), January 24, 1978, p. 5; and "Defence Production," *The Overseas Hindustan Times*, February 16, 1978, p. 6.
18. Ibid.

19. Ibid.
20. Ibid.
21. Derek J. de Solla Price, "The Implications of Theoretical Studies for Decision-Making in R and D Management," in OECD, *Management of Research and Development* (Paris: Organization for Economic Cooperation and Development, 1972), p. 276.
22. National Committee on Science and Technology, Government of India, *Draft Science and Technology Plan 1974-79* (New Delhi: Department of Science and Technology, Government of India, 1974), p. 3.
23. Press Information Bureau (Defence Wing), Government of India, "Defense Research and Development Organization of India." Press release, 1978; and, Major-General C. Sundaram, "Role of Technical Committees in Indigenisation of Defence Stores." Press release, 1978.
24. Stockholm International Peace Research Institute (SIPRI), *Resources Devoted to Military Research and Development* (Stockholm: SIPRI, 1972), p. 77; and from interviews with officials of India's Defense Research and Development Organization.
25. Kavic, *India's Quest for Security*, p. 136.
26. "Third World Could Make Cruise Missiles," *Christian Science Monitor*, August 4, 1977, p. 13.
27. *Jane's Fighting Ships: 1976/77* (New York: Franklin Watts, Inc., 1976), p. 719.
28. "India Reported in Position to Make IRBM Warheads," *India Abroad* (New York), November 10, 1978, p. 1.
29. See, P. R. Chari, "The Jaguar Aircraft Deal," *India Abroad* (New York), October 27, 1978, p. 2; and, "French Offer," *News and Cine India* (New York), September 29, 1978, p. 1.
30. "Indian Naval Developments Foreseen," *International Defense Review* 2 (1978):147-48.
31. Collated from: (1) *SIPRI Yearbook, 1974* (Cambridge, Mass.: MIT Press, 1974), p. 215; (2) United States Arms Control and Disarmament Agency, *World Military Expenditures and Arms Transfers, 1965-1974* (Washington, D.C.: U.S. Government Printing Office, 1976), p. 32; (3) *The Military Balance, 1978-1979* (London: International Institute for Strategic Studies, 1979).
32. Ruth Leger Sivard, *World Military and Social Expenditures 1977* (Leesburg, Virginia: World Military and Social Expenditures Publications, 1977), p. 26.
33. Emile Benoit, *Defense and Economic Growth in Developing Countries* (Lexington, Mass.: Lexington Books, 1973), especially

chapter 4, "Defense and Development in India," pp. 149-220.

34. Ibid. According to Benoit, "India is a country that contributes to our general finding of a positive correlation between the burden of defense expenditure and the rate of growth," p. 126.

35. United States Arms Control and Disarmament Agency, *World Military Expenditures*, p. 62.

36. Ravi Rikhye, "India Continues Military Buildup," *Armed Forces Journal*, August 1972, p. 29. According to the author, "Informed sources say that Indian military strength is greater than India admits. These sources say that India has at least 33+ division equivalents, whereas only 27 are on the books. The Air Force and Navy appear similarly underrated."

37. See *The Military Balance 1978-1979*, p. 57. Of China's 5,000 combat aircraft, 4,000 are MiG-17s/-19s. It has only 80 MiG-21s and "some F-9 fighters."

38. See, Raju G. C. Thomas, *The Defense of India: A Budgetary Perspective of Strategy and Politics* (Delhi: Macmillan Company of India, 1978), p. 147, Table 4.

39. Joel Larus, "The Indian Navy: The Neglected Service Expands, Modernizes, and Faces the Future," draft paper, 1978; Vice-Admiral N. P. Dutta, "Growth of Shipbuilding Industry in India," Press Information Bureau (Defence Wing), Government of India, press release, 1978; and, *Far Eastern Economic Review*, June 3, 1974, p. 30.

40. See, "Blasts That Aren't Barred," *The Overseas Hindustan Times*, August 3, 1978, pp. 1, 13.

41. Ibid.

42. While China has nuclear weapons, it does not possess a single commercial nuclear power station. It recently contracted with Framatome of France for the construction of its first such facility.

43. *The Military Balance 1978-1979*, p. 5. The missile, apparently, is "unlikely to become operational for some years yet."

44. Department of Space, Government of India, *Annual Report 1977/78* (Bangalore, India: Indian Space Research Organization, 1978).

45. See, K. S. Jayaraman, "India's March Into Space," *Indian and Foreign Review* (Delhi) 16 (March 1, 1979):15-19; and, "Development of Space Technology," *India News* (Embassy of India, Washington, D.C.), April 10, 1978, p. 1.

46. For an analysis, see Onkar Marwah, "India's Perspectives on U.S.-Soviet Naval Rivalry in the Indian Ocean," in George Quester, ed., *Naval Arms Control* (Aspen, Colorado: Aspen Institute for

Humanistic Studies, forthcoming).

47. Richard Burt, "U.S. Weighing Added Sale of Arms in Persian Gulf," *The New York Times*, March 7, 1979, p. 2.

48. For a review of the problem, see Onkar Marwah, "Northeastern India: New Delhi Confronts The Insurgents," *Orbis* 21 (Summer 1977):353-73.

49. India's west coast is subject to heavy monsoon squalls, while cyclones occur frequently along its east coast. So, the cost of the radar system is probably written off against disaster-relief expenditure.

50. "Future Plans for Indian Coast Guard," *International Defense Review* 1 (1979):126; and, "A Coast Guard for India," *India News* (Embassy of India, Washington, D.C.), August 21, 1978, p. 5.

5
Japan's Security Perceptions and Military Needs

Yasuhisa Nakada

Introduction

During the three decades since World War II, Japan has built a modern, prosperous, consumer-oriented society that feels no real sense of threat by either military attack or military pressure. The principal objectives of Japanese policy have been to assure, first, that Japan becomes and remains a major economic power and, second, that Japan carefully avoids taking any politico-military responsibility beyond the nation's boundaries. A long-time political leader of the Liberal Democratic party has deftly argued that Japan's strategy should be an *amoral* diplomacy that appears to make no attempts to assume any international burden.

However, through relations with the United States, the Soviet Union, and China, the Japanese have acquired an awareness of the growing necessity for independence if Japan is to remain a legitimate sovereign state and a major Asian power in the 1980s. Factors contributing to the birth of such an awareness include hard bargaining between the United States and Japan over the steel issue, criticism over trade surpluses, the sharpest rise in the value of the yen since World War II,

---

Funding for this research was provided by a Ford Foundation Fellowship for Mid-Career Journalists, undertaken at the Harvard University Program for Science and International Affairs (PSIA) during the spring of 1977, and a 1977-1978 Fulbright award at the Harvard University East Asian Research Center. I am also grateful to many individuals at PSIA who commented on my work and in particular to Jonathan Pollack and Onkar Marwah, who encouraged me to write this essay.

difficult negotiations about Japanese nuclear independence from the United States, and public recognition of the vast military power of the Soviet Union, which presents a potential threat to Japan's resources, energy, food, and national security.

All of these factors have contributed to an awareness of the need to establish an integrated policy by and for Japan. Increasing numbers of political, economic, military, and media representatives realize Japan must reassert herself and her international responsibility, thereby beginning to pursue positive international goals. My experiences as a reporter covering the prime minister's office, the Foreign Ministry, and the Defense Agency during the past five years strongly confirm this observation. Another option has begun to emerge in the minds of Japanese politico-military policymakers for the 1980s that differs from the above-mentioned amoral diplomacy. Growing numbers of Japanese leaders want to establish a policy under which Japan will be operating through a system of alliances and various international arrangements with foreign countries. This chapter will discuss these new initiatives, focusing on present and prospective policy frameworks pertaining to national security as defined by concerned members of the politico-military establishment in Japan.

It is my principal contention that Japan will one day acquire a global political identity—comparable to her present economic identity—that will render her a great independent power free to pursue national objectives and goals. In addition, Japan is now creating, in case of need, a modern, diversified military force. However, these topics remain highly delicate in Japan. Until recently, public discussion of security and defense issues has been largely taboo. There is still a conscious effort in many circles to avoid all public debate of security and defense problems. Regrettably, this means that the subject of security affairs is still not entirely respectable for scholars, government officials, and journalists. Thus, although Japan first rearmed in 1954, there has been an almost total lack of analysis of the activities of Japan's Self-Defense Forces by scholars as well as the public. The primary task at this stage, therefore, is to provide some basic facts and political perspective as groundwork for better understanding and further research.

This chapter will discuss five basic issues: (1) Korean security and how it is perceived by the Japanese politico-military elite; (2) the new Asian political and strategic environment; (3) the formation of a national self-consciousness; (4) the basic objectives of politico-military strategy in Pacific Asia in the 1980s; and (5) the probable configuration of power and policy that Japan will assume. Much of the necessary evidence for such research is difficult to acquire. However, by specifically interviewing many involved with security questions, it proved possible to establish contacts with hundreds of politicians, economic leaders, governmental officials, and uniformed officers. In addition, sensitive official materials were made available as well as open literature closely connected with national security matters.

Japan at the Crossroads

Withdrawal of U.S. Troops from South Korea

Japan today faces a new security phase. The most important and variable element lies in the future of the Korean peninsula, an area that significantly affects Japan's security.[1] Early in his administration, President Carter had pledged to withdraw the remaining U.S. ground forces from the Republic of Korea over a four- to five-year period. He subsequently dispatched General George Brown, then chairman of the Joint Chiefs of Staff, and Philip Habib, then undersecretary of state for political affairs, to meet with President Park and then Prime Minister Fukuda in order to express U.S. views directly. According to Habib, "The Japanese government conveyed its concern that the ground forces withdrawal be carried out in an appropriate manner which would not endanger the security of the Republic of Korea nor threaten the security of Northeast Asia."[2] In addition, U.S. officials pledged close consultations with Japan and the South Korean government on this withdrawal question.[3]

Yet the unilateral nature of the proposed troop withdrawal directly contradicted the latter commitment. Indeed there was an overall absence of government review on the issue within the United States as well.[4] Why did Carter hasten the withdrawal?

No doubt his decision was linked to his determination to reorganize U.S. foreign policy and to uphold various campaign pledges. Other factors, however, were clearly involved. There was undeniable evidence of an alteration in U.S. policy toward Asia. This change was expressed by Paul Warnke, former director of the U.S. Arms Control and Disarmament Agency, who argued that it would be quite appropriate for the United States to withdraw ground forces from Korea in five years.[5] Warnke had conveyed similar views in an article published shortly before assuming office:

> The continuation of an independent South Korea has a bearing on Japanese perceptions of their own security. In considerable part, however, the importance of South Korea to Japan's security is a product of our constant repetition of the theme to the Japanese audience. As with the once-perceived threat from Red China, it will likely dim with time unless kept polished by American rhetoric. Here again, the loss of a mutual enemy should not cost us a friend.[6]

Such a view, however, has paid inadequate heed to a host of unknown factors and variable elements in Japan and in the region.[7]

Faced with the Carter policy, the Japanese establishment feels discomfort and anxiety because the nation has had close, strong ties with the Korean peninsula—historically, geographically, politically, and militarily. The decision to withdraw troops indeed had a psychological impact on Japan. Concern seemed particularly appropriate in terms of a major U.S. interagency military assessment that concluded that North Korea would have advantages over the South "in all categories at early stages of a renewed war."[8]

In addition, many in Japan feel that the situation in Northeast Asia remains unstable. What especially vexes them is that the United States and South Korea are going in divergent directions ideologically and diplomatically. It is of the utmost importance to them that the United States thinks of its military interests in Korea apart from events such as the Korean influence-buying scandal.

Undeniably, however, the projected U.S. troop withdrawal

seemed in part the result of recent U.S.–South Korean tensions, which were created not only by the U.S. Congress bribery scandals, but also by the plight of political prisoners in South Korea and the authoritarian regime of President Park Chung Hee. In my interview with Congressman Les Aspin, he made it clear that rumors of bribery in the Congress had created an undeniable tension. As far as congressional members were concerned, they wanted no part of the Korean troubles.[9] As for the human rights issue, the Japanese establishment's concern is that the crisis-oriented reporting in the United States might transform and reduce the *security* issue into a morality issue. Aside from the political and military implications of the withdrawal, many Japanese feel that an "Asian gook syndrome" prevails in the United States. During the Vietnam war, many commentators and editors created an impression that the South Vietnamese were not worth assisting.[10]

Japanese decision makers, then, prefer former Secretary of State Kissinger's stress on peace and security in the Korean peninsula rather than the human rights issue. Many view the general U.S. attitude toward the Korean problem as one of "crusaders" against the South Korean regime. The Carter administration and an overwhelming Democratic majority in the Congress, these individuals contend, will prompt an initiative against South Korea sooner or later. Thus, the Carter administration viewed the human rights campaign as a means of gaining public support for a new, more idealistic conception of diplomacy. In view of such opinions, numerous Japanese fear that the United States will try to formulate a new Korean policy in the next four years and leave the Asian scene.

To be sure, the State Department has tried to keep the security and human rights issues separate. Secretary of State Cyrus Vance has told a Senate subcommittee that because of overriding security commitments, the United States would not reduce its aid to South Korea no matter what their violations of human rights.[11] President Carter has further stated that any slowdown in U.S. military deliveries to South Korea would upset the timetable for the scheduled withdrawal. Any stalemate in the program of U.S. security assistance to South Korea would represent a danger militarily because a withdrawal of U.S. troops would still be conducted without a quid pro quo.

In communication, the Japanese often attach more importance to that which is unspoken than to the explicit. Thus, it is particularly notable that high-ranking politico-military officials in Japan have stated publicly that Japan opposes the withdrawal of U.S. troops from South Korea. The government wants not only continued maintenance of U.S. troops, but also wants no abrupt change of the U.S. Korean policy. Yet they predict the Carter administration will complete its scheduled withdrawal in 1982, despite recent indications to the contrary, a move obviously at variance with the Japanese government's desire that the United States keep ground forces in Korea.

What are the specific military implications of a withdrawal of troops from South Korea? The major reason why the United States wants to withdraw ground troops is to reduce the risk of automatic involvement in another land war in Asia. Under such circumstances, the negative mood toward automatic involvement in a war will not weaken but rather will become stronger. Obviously, the present South Korean government is not in the good offices of the United States. This has been clearly indicated by Henry Owen: "In the short term, our purpose must be to avert a Korean conflict. Over the longer term, we should try to distance ourselves from events in Korea, and to render the U.S.-Japanese connection less vulnerable to events on the peninsula."[12]

Despite such sentiments, Japanese military experts evaluate the withdrawal of troops on a radically different premise. U.S. ground forces in Korea are viewed as a critical balancer of the region; hence, there is no specific reason for an abrupt withdrawal. And, although the Carter administration feels this problem should be resolved in close consultation with Japan, there is no evidence that such advance consultation was ever undertaken. Maintaining an air force in Korea is viewed simply as a token pose toward Japan, the true intention being to maintain an air force until there is a basic change in the Korean situation or in Japanese attitudes.

U.S. policy thus reflects an overall movement away from the principle of forward-based deployment in Asia. As recently as

1976, a statement from the U.S. Department of Defense defined U.S. policy as "maintaining two principal strong deployments . . . in central Europe and northeast Asia and being able, in conjunction with allies, to hold a forward defense line against a major attack in either theater."[13] Now, however, U.S. officials seemingly regard the army's Second Division as militarily unnecessary and see its continued presence near the thirty-eighth parallel as a commitment to instant and automatic involvement in any renewed warfare. Similarly, while the air force wing is perceived as necessary to compensate for the supposed weakness of South Korean air power, the United States wants to deploy the air wing somewhat further south on the peninsula, away from the thirty-eighth parallel.[14]

Japanese military experts, however, feel that maintenance of the present U.S. Army Second Division and the Air Force Fighter Wing in Korea is absolutely essential. The disparity between U.S. political intentions and Japanese politico-military affairs has thus increased sharply. Contributing further to Japanese discontent is the fact that no explanation has come forth as to why the U.S. government may change the "trip wire" system established in the 1950s.[15] Thus, the Japanese Defense Agency cannot understand why the Forward Based System (FBS) concept will be changed.[16] These sources also fail to comprehend why the United States can withdraw its ground troops when the U.S. military presence helps to deter North Korean attack. The Japanese military establishment also believes that the U.S. forces in Korea have strategic value in any holding action against either the Soviet Union, China, or both. In short, the United States has not yet crystallized any logic for withdrawal that proves satisfactory to Japan.

The change in the international concept of U.S. troops as a balancer implies strategic change in the concerned countries.[17] If the withdrawal is carried out, it could well set off a trigger reaction in the status quo in East Asia. Among the Japanese military establishment especially, there is a growing awareness of an eventual U.S. withdrawal, because the Korean issue is viewed as a potential catalyst for the remilitarization of Japan. The most striking opinion about Japanese counteraction has been voiced by Mr. Maruyama, administrative vice-minister of the Japanese

Defense Agency. He stated that should U.S. troops withdraw from the Reublic of Korea in its present situation, the basic premise for the formulation of a "standard Japanese Defense Force" would become totally invalid. He suggested that the Japanese government would have to review its entire new Defense Force plan. He also indicated that necessary steps will be taken to expand military capability in the future. This mood calling for revision is, as a countermeasure to the Korean situation, leading to actual preparedness in Japan's Fourth Defense Buildup.[18]

The central point in Maruyama's remarks was that the presence of U.S. troops in Korea formed the basic premise of Japan's defense concept. He hinted that the Japanese government would review the whole Japanese defense buildup plan and also suggested a considerable expansion of military capability in the future. Therefore, it is essential to consider the looming impact of the withdrawal of U.S. ground troops from South Korea as the most important question in Japan's future security. No area is more important to Japan's peace and security than Northeast Asia. And it is in this region that the interests of the United States, China, and the Soviet Union converge and conflict in a complicated pattern.[19]

Japan's Gravest Era

At the end of November 1977 Prime Minister Fukuda had carried out a sweeping reorganization of the cabinet, further stating that Japan faced the gravest situation since World War II.[20] Foremost among these concerns are economic issues. There are dangers of a backlash in the present trade confrontation between Japan and the United States. U.S. pressure for economic concessions has created strong counterarguments in Japan.[21] The reality is that no handful of actions reached at a bargaining table will suddenly or drastically change the entrenched national economic system reflecting Japan's insular structure. Yet the Carter administration has warned Japan that it must rapidly eliminate its giant trade surplus with the United States lest domestic pressures for protection against Japanese imports become extensive.[22] The United States feels that the Japanese have an obligation to the world system to stimulate their economy so Japanese will buy more from abroad and make greater efforts to relax tariffs, quotas, discriminatory regulations, and other barriers to imports, because concessions there will help pacify powerful factions in Congress.

Why is Japan unable to cut its surplus? For Japan, doing more than has been offered in negotiations with the United States requires not only government decisions to spur growth but also changes in entrenched habits among Japanese industrialists and consumers. Domestic demand simply is too weak to provide the needed thrust to attain the targeted economic growth. Therefore, massive quantities of goods flow into foreign markets, particularly in the United States and Europe. To make matters even more complicated, domestic political pressures prevent Japanese leaders from going much farther at present. They say the present liberalization measures are the best they can do because of fierce domestic opposition, especially among farmers who represent many votes to the ruling Liberal Democratic party (LDP). In the course of this potential "trade war," there is a growing feeling among Japanese that legitimate, long-term policies must be established. Such needs must be realized in order for Japan to establish its position and relationship with the outside world.

At the same time, Prime Minister Fukuda had stated that the United States was largely to blame for the current worldwide fluctuations in money markets and should make more efforts to maintain the value of the dollar at an appropriate level.[23] Some officials further argue that Japan cannot be expected to reduce or even eliminate trading surpluses when the nation is almost totally lacking in fossil energy reserves. Thus, large trade surpluses are regarded as a needed cushion against an uncertain future. What is more important, future policy options are based on the belief that: (1) current frictions with other countries basically grow out of Japan's ingenuity and hard work;[24] (2) a more significant indicator is a country's overall accounts, which involve the total exchange of goods and services (for example, the balance of exports and imports) rather than any bilateral trade balance; and (3) the export-oriented economic structure will be indispensable to Japan's prosperity in the future.

While it is difficult to find an effective alternative, Japanese opinion has begun to react against dictated U.S. policies. Many now contend that the United States does not fully appreciate what Japan has already done and is simply looking for a scapegoat to deflect attention from its own inability to control energy imports. If the United States pushes too hard it may set off in Japan a vehement nationalistic reaction. Beyond the next few years, Japan's concern will center on the expanded economic battle between Japan and the United States and how to cope with it.

Besides the necessity for hard trade negotiations between Japan and the United States, the Japanese also feel domestic integration and solidarity are essential before confronting divisive diplomatic issues, such as the Japanese-U.S. nuclear negotiations, the Japanese-Soviet fishery negotiations, and the U.S. determination to withdraw troops from Korea. A mood of foreboding and even national peril has begun to pervade leadership thinking, bringing with it a readiness to reassess the deep aversion to securing and protecting Japanese interests which characterized defense policy following World War II. The media now argues about how best to find and define the national purpose of Japan. All such trends spell the end to the noncommittal attitude heretofore present in public thinking.[25]

Japan's attitudes toward foreign countries have long been passive and egocentric, partly because of the legacy of the U.S. occupation and partly because of reverse self-satisfaction as the non-Communist world's second largest economic power. However, the Japanese establishment has recently undergone an awakening through changes in the international climate. They realize Japan can no longer play a limited role overridden by memories of war stigmas.

Yet it is also true that conservative decision makers within the Japanese government—especially the pro-U.S. camp, represented by the Foreign Ministry—do not want to admit officially that Japan stands at a crossroads. This camp still tries to maintain the same policy line, content with the status quo. The crux of their interests lies in their concern with what the world can do for Japan and not what Japan can do for the world.[26] In the changing international situation, however, Japanese foreign policy will have to change. As Nathaniel

Thayer has noted, Japanese diplomacy now appears willing to take the political initiative instead of waiting to respond to events.[27] In what specific policy areas can evidence of these changes be seen?

Areas of Flux

The United States' Changing Posture

While Japan has begun reacting to external and internal political pressures, U.S. leverage and influence on her allies in Pacific Asia are declining. Perhaps it is fairer to say that the Japanese politico-military establishment believes that the United States will formulate a new Asian policy, namely, an "off-shore" strategy, which some U.S. policymakers have already suggested.[28] Yet, as Jerome Cohen points out, it is quite difficult to state as declared policy what the future U.S. Asian position will be.[29]

Future U.S. policy toward Asia, Japanese experts believe, will most likely be influenced by a drop in U.S. possessiveness about protecting Asia. There are no U.S. ground troops anywhere on the Asian continent except in South Korea. The U.S. position in Pacific Asia is more favorable than it has been since the end of World War II. Thus, the Japanese security establishment realizes fully that most Americans feel the era of U.S. military commitment on the Asian mainland is over.

But can the United States really keep a safe distance from the Asian mainland? In terms of strategy, the Carter administration appears headed in the direction of the Nixon Doctrine, which called on Asian nations to provide their own first line of defense and for the United States to support that line from the sea and air.

Yet the disengagement from Southeast Asia, the planned troop withdrawal from South Korea, and the growing willingness to cut formal ties with Taiwan suggest that the United States is less interested in defending Asia. Gnawing doubts about the reliability of U.S. defense commitments to Pacific Asia have developed. Concerned Japanese perceive that Americans interpret Asian involvement as an endless commitment to countries of secondary importance and to causes of questionable relevance to U.S. security.

Asia since World War II has been an area for "hot war," which erupted under the antagonism of cold war tension between the United States and the USSR. Asia was the arena of intermittent limited war for twenty years, most notably in Korea and Vietnam. And, notwithstanding renewed assurances about U.S. commitment to the defense of South Korea, the withdrawal of all remaining ground troops from the Asian continent will have great impact on perceptions of U.S. policy. It would imply the accomplishment of military disengagement from the Asian mainland and reversion to the traditional U.S. strategy toward Asia that prevailed during the Russo-Japanese war era.

Similarly, the Maritime Self-Defense Forces (MSDF) are concerned about the continuance of U.S. bases in the Philippines. A reduction of base areas under U.S. control and their transfer to Philippine jurisdiction will be perceived as destabilizing security in the western Pacific, where vital sea lines of communication are located. The MSDF is concerned about control over these territorial waters. These officials feel a tentative agreement on the bases pact means a changing U.S. posture. They fear such changes will create new uncertainties in the maintenance of security and peace in the area, as the United States has long had dominant influence over the Philippines. After the U.S. defeat in Vietnam, doubt about the credibility and commitment of the United States has increased. The enhanced powers of the Congress to limit the president's freedom of action in foreign affairs has also been recognized. Maritime officers further feel that as the U.S. military presence in Asia has shrunk, U.S. interests have changed. At the same time, the United States has tried to urge Japan to take more security measures in the western Pacific area. Such moves suggest that the ultimate objective of U.S. policy toward Japan is aimed at her becoming an anchored aircraft carrier and supply ship at the center of the U.S. system of Asian alliances.[30]

Various Western nations and some Asian countries are also urging Japan to take steps to assume greater international responsibilities—particularly within her own region—both economic and political. For example, one new trend was evidenced by Prime Minister Fukuda's trip to the members of the Association of Southeast Asian Nations (ASEAN), during which he pledged $1 billion in assistance to help bolster the economies of the region.[31]

Without question, therefore, the equilibrium that has largely dominated postwar Asia appears to be breaking down. As Kiichi Miyazawa suggests, the future relationship between the United States and Japan will be marked by a tacit trade-off.[32] An important determinant of the future U.S.-Japanese relationship will be how both governments deal with one another on major policy issues that arise between them. Japan, as a competitive but cooperative nation, will share economic and political burdens in Asia with the United States, thus establishing a genuine partnership between the two countries. This new move may impinge upon the U.S. position, but it is the necessary result of multipolarization. There is no real partnership as long as Japan is confined as a lonely economic giant in the Far East.

If the "lonely giant" mentality is no longer an adequate position for Japan, what types of policies might be required to replace it? It is in this area that Japan's dramatically changing relationship with China reveals a good deal about Japanese thinking.

Japan and China

A new axis between Japan and China will enhance Japan's strategic importance during the 1980s for other Asian states. The successful conclusion and signing of a Treaty of Peace and Friendship in August 1978 is the most obvious and significant embodiment of this rapidly emerging relationship. This agreement, however, does more than solidify the Sino-Japanese relationship that has developed since the normalization of relations in September 1972. It testifies to Japan's capacity for political and diplomatic independence.

Until recently, Japanese diplomacy has been bound and dominated for years by the postwar U.S.-Japanese alliance. Political and military officials have long awaited and anticipated room for independent strategic options in conducting legitimate foreign policies, a goal that the new treaty with China will help establish. The consensus is that Japan's new policy needs and her political alliance with China have become necessary due to vital political-military interests as well as particular resources and energy problems in Asia. Japanese policymakers have concluded that the new regime led by Hua Kuo-feng will remain stable and pragmatic into the 1980s. They also regard China as a country that is more cooperative toward Japan than many other nations.

As a result of the recent trade war with the United States, many Japanese predict that China will become an increasingly vital trading partner for Japanese industries. In 1975, trade between the two countries reached $3.79 billion, the largest share of China's foreign trade with a single nation. Japan's imports of Chinese oil also increased sharply, rising to about 8 million tons. Increasingly frequent exchanges of economic trade missions have also been evident. And in February 1978, Japan and China concluded a multiyear $20 billion trade agreement. China will supply crude oil and coal in return for Japanese exports of steel, industrial plants, and technology. Thus in the 1980s Japan and China will become more and more complementary to each other in politico-economic-military terms—witness China's eagerness to learn from other nations that have sanctioned and stimulated foreign trade, arms purchases from the West, and the industrial development goals now being implemented.

The Japan-China relationship should also provide Japan with valuable leverage in the delicate balance among superpowers in Asia. Proponents of the treaty within Japan have long argued that a decisive attitude by Japan to conclude the agreement would be a positive move in helping abrogate the thirty-year Sino-Soviet Treaty of Friendship, Alliance and Mutual Assistance, aimed specifically against Japan. The continued existence of this 1950 treaty presents a dilemma for Japan. Tokyo might face retaliatory measures by the USSR, since the new treaty contains what the Soviet Union regards as an antihegemony clause aimed at the USSR. The Japanese explanation is that the peace treaty is strictly a matter between Tokyo and Peking. Japan has also sought to counter any possible charge that a new anti-Soviet Peking-Tokyo axis is being set up on the grounds that "antihegemonism would not be aimed at a third nation, would not mean a military alliance, would be applicable to regions outside of Asia as well, and would be in line with the United Nations Charter."[33] The fact that the agreement does include an article stipulating that "the present treaty shall not affect the position of either contracting party regarding its relations with third countries" gives added weight to Japanese arguments. Such Sino-Japanese convergence is largely attributable to the declining leverage and influence the United States has on her Pacific Asian allies. The Chinese are now perceived as favoring a strong U.S. presence in Asia as insurance against the expansion of Soviet influence. They supposedly consider a U.S. withdrawal as regrettable. Some analysts even suggest that China opposes the phaseout of U.S. military forces from Korea, since Peking fears it would result in increased Soviet influence in the Far East and finally destabilization of the status quo. According to this view, Chinese officials are particularly uneasy about indications of both a Soviet naval presence near the China coast and a larger Soviet naval presence in the Pacific Ocean. Attitudes of both Japanese and Chinese defense planners toward Soviet power in the Pacific clearly have much in common.

Thus, Northeast Asia is the arena where both Western and Communist superpowers interlock. For instance, from the Chinese point of view Northeast Asia, especially the Korean peninsula, is a buffer zone against any Soviet approach into the Yellow Sea. Soviet military growth in the Far East is clearly of great concern to both China and Japan. In addition to the deployment of strategic weapon systems, the USSR has greatly built up its general purpose forces. Moreover, the growth in the strength of Soviet forces in Northeast Asia has been at the precise time when the United States has shown an inclination to withdraw from the Asian mainland and concentrate on the defense of Europe.[34]

Although not all Soviet power is directed against Japan, defense planners in Japan must still find ways of managing and integrating ties with a triangle composed of the United States, USSR, and China. The Soviets have also been expanding their naval forces, which include the world's largest submarine fleet consisting of about 250 attack submarines.[35] For the present, it can be safely projected that the Japanese will continue to seek a U.S. security commitment in the region, including further defense cooperation between Japan and the United States. Japan thus still depends heavily on the Japanese-U.S. security treaty. Most Japanese believe the bilateral defense pact contributes to the nation's peace and security. But this heavy dependence on the United States has dulled the sense of self-reliance.

China no longer objects to such security links. In addition, Peking has made unusual overtures and displayed a cooperative attitude toward Japan. Japanese politico-military circles interpret the current Chinese policy change as deriving mainly from the need for military cooperation with Japan. In an apparent policy shift, the Chinese government has invited high officials of Japan's Self-Defense Forces to her soil. The subsequent visit by military experts was the first the Japanese government had officially sanctioned since World War II. Japanese officials interpret such invitations as a Chinese response to the future withdrawal of U.S. troops and to the Soviet military buildup in the Far East.[36]

China is launching a buildup of its own forces, as well. In the process of modernizing military capabilities, the Chinese establishment faces a serious lack of capital, machinery, tools, and the know-how necessary to create a sophisticated fighting force. Chinese leaders would like aid from advanced U.S. technology but the Chinese are also looking to other sources, including Japan. After consultations, Japanese military experts have the strong impression Chinese leaders long for close cooperation with Japan in the promotion of industrial modernization, specifically heavy industry and technology, including radar technology, ship building, tank and armored carrier design, and computer know-how. There is a growing awareness of competition in arms sales or arms-related sales to China through Japan, the United States, and other nations. The Chinese government is showing interest in purchasing Western weapons that will help modernize its army.[37]

Japan's primary positive contribution to Chinese strategic

capabilities is in economic areas. It is no understatement to describe Japan as the vehicle for Chinese strategic and military development, a function the Soviet Union once fulfilled but which Japan is rapidly assuming. The constituency-building potential of arms sales at large to China is quite a promising field for Japan. There are legitimate technological requirements that Japan might fill for China, thus establishing a complementary relationship between the two nations. Thus, Japan and China are cooperating increasingly with each other through a variety of political, diplomatic, and economic arrangements. Such developments suggest new directions in Japanese policy that until recently would not have proven possible. How have such trends been reflected within Japan?

New Directions

Reformation and Consolidation of Japan's Politico-Military Structure

While the Japanese situation is now somewhat unclear and uncertain, Japan's strategic changes can be described under the following categories. Reformation and consolidation are reflected in a number of areas, each of which will now be briefly discussed.

Public attitudes toward the Self-Defense Forces. A public opinion poll on the Japanese Self-Defense Forces (SDF) released by the Defense Agency on October 30, 1977 (the SDF's twenty-third anniversary) revealed that more than four out of five Japanese approved the existence of the SDF. This was the first time that popular backing had reached this level.[38] This figure was a reaction in part to the Communist victory in Indochina and the MiG-25 incident (in which a Soviet pilot defected at Hokkaido). These two incidents augmented public concern over Japan's security and also stirred opinions about the strategic importance of the Korean peninsula and the severity of the situation surrounding northern Japan.

In this regard, the Soviet Union has assumed a tough posture towards Japan on several issues. It has refused to let Japanese fishermen into its newly claimed 200-mile fishing zone. Talks about the return of four northern islands seized by the USSR in World War II have yielded no results. And Soviet warships and reconnaissance aircraft have been sent close to Japan to underline a hostile presence.

The security situation is perhaps most critical in the naval area. Japan has three major international straits used by both surface and undersea craft of many nations; some of these vessels presumably carry nuclear weapons. These three straits are the Soya (or La Perouse) Strait north of Hokkaido, the Tsugaru Strait, and the Tsushima Strait in the southern part of the Sea of Japan. Soviet naval activity has been linked to transit through these locations. The blue water capacity of the Soviet Pacific Fleet has been progressively strengthened. The USSR's naval forces have been building support, oiling, and landing vessels in its continuing effort to upgrade blue water capacities. Its wide-ranging ocean exercise in 1975, OKEAN 75, aroused strong Japanese concern.[39]

The growing Soviet military presence is arousing concern not only about a conventional military threat per se, but rather a combination of power, political influence, and the possible use of force. Japanese officials realize that this military buildup in the Pacific is not military power for destruction. Rather, it is a demonstration of resolve and justification of national interests against political pressure from other nations. Thus, Japanese defense specialists feel that the United States must maintain a powerful naval and air presence in the northwestern Pacific and ground forces in the Korean peninsula, which can be reinforced as necessary to deter Soviet involvement in any future conflict.

The government and public recognize the overwhelming military superiority of the Soviet Union vis-à-vis Japan. The above incidents heighten Japanese frustration over the inadequacy of their self-defense mechanism, as reflected in the public opinion poll. The Defense Agency regards this change in popular support as important because in the event of a crisis—either external aggression or internal subversion—the degree and nature of public support will be decisive.

In addition, heightened public awareness of security issues is required in view of continuing threats to Japanese access to resources. Fishing disputes with the Soviet Union in the northern islands can be cited, as well as seabed oil development in the East China Sea shared with China, North and South Korea, and Taiwan. Thus, increasing public knowledge and awareness of security issues are deemed necessary because such consciousness can counter the so-called coercive diplomacy of bilateral negotiations.

The role of the new Special Committee on Defense in the Diet. New security forums in the parliamentary and administrative complex are considered indispensable vehicles in fostering healthy nationalism. In May 1977 the Diet approved establishment of a Special Committee on Defense to discuss national security issues. Such a forum at the domestic political level meant a breakthrough in fostering public awareness of security issues. As Seiichiro Ohnishi has pointed out, this step has clear significance for political trends within Japan: "Such an atmosphere aroused the legislature from its lethargy regarding national defense. . . . Indeed, it had been 25 years since a revision law was first proposed by the Progressive Reformist Party, at the time of the establishment of the National Safety Agency in 1952, designed to establish such a committee."[40]

The raison d'être of this new committee lies in its becoming a forum on national policies, affording a place where the ruling party and the opposition parties can debate various security problems on an equal basis. This has not been possible as the LDP has, since World War II, occupied itself primarily with defending its own policies while opposing parties never engaged in serious debate, but merely opposed the LDP for opposition's sake.

Self-Defense Forces' strategic estimate. A significant change has developed in the attitude of civilians and uniformed officers of the Self-Defense Forces (SDF). They are openly positive about coping with a new Asian situation and have called for the government to review the total plan of the "Standard Defense Force" and basic concepts behind it. Especially since President Carter expressed a determination to withdraw ground troops from Korea, the attitude of uniformed officers has become very positive. In May 1977 top-ranking officials of the Self-Defense Forces, in a rare meeting with Prime Minister Fukuda, proposed that the government review the total plan and basic principles of the "Standard Defense Force." The gist of this meeting revealed that the Self-Defense Forces have a different view of Carter's decision than the government, which has publicly accepted U.S. pronouncements. Strikingly, a few days later Fukuda personally attended a meeting of the Self-Defense Forces in Tokyo and pledged his support for their mission.

The most likely strategic measure in the 1980s will be a response to the Korean question. The Defense Agency's evaluation of the possible withdrawal of American troops from South Korea changed to a great extent between 1976 and 1978. In early 1976, Takuya Kubo, former administrative vice-minister of the Defense Agency (and later secretary-general of the National Defense Council), made it clear that U.S. forces in Asia, specifically those in South Korea, had an indispensable balancer role in the Far East, helping to maintain stability there. He also mentioned assurances from the Pentagon that the United States intended to maintain forces there.[41] At the end of the year, after Carter assumed office, Kubo opposed the nonchalant way in which U.S. withdrawal plans were presented. As a spokesman for the Japanese government, he said that Japan wants prudent, careful action from the United States in dealing with this question.[42] The reason is clear: Japanese defense authorities, in the final analysis, feel the Korean peninsula is vital to Japan's defense.

Yet another problem is the relationship of any changes in the Korean situation to Japan's overall defense plan. The Japanese government has formulated a National Defense Program Outline that is intended to serve as the fundamental document guiding the development, maintenance, and operation of the defense structure.[43] Any change in the status quo on the Korean peninsula would require careful review and revision of such plans.

After recent developments, planners within the Defense Agency have pushed for establishment of a security posture that can cope with any major strategic changes. This new mood has led to improvements in equipment, modernization of command and control systems, and building up logistic support for the SDF.

Self-Defense Forces reorganization. The Self-Defense Forces are planning to establish a central command post within the next five years, consolidating the Air Defense System and the Naval Defense Complex into one strong, effective organ. This move is considered the first step toward formulating the core of a general staff for the first time since World War II.[44] At a symbolic level, General Hiroomi Kurisu, chairman of the Joint (Chiefs of) Staff Council asked Prime Minister Fukuda to create an opportunity for top SDF leaders to report strategic briefings to the prime minister periodically and to sign in at the Imperial Palace at governmental ceremonies as a public demonstration of the SDF's having standing equal to the vice-ministerial level.[45]

The Japanese are also now involved for the first time in the early stages of joint contingency planning with U.S. forces based in their own country in order to coordinate responses should Japan be attacked. Some sectors acknowledge that Japan should not depend forever on the United States for every aspect of national security planning.

Another possibility Japanese defense leaders anticipate is strengthening strategic ties with the United States and South Korea, mainly by beefing up the information exchange of early warning systems. They want to expand the strategic intelligence network throughout the Far East. The Defense Agency also points out there will be new developments in the buildup of logistics support capability and standardization of military weapons among the United States, Japan, and South Korea.

Strategic preparedness after fourth defense buildup. In 1957, Japan initiated the first of a series of five-year defense buildup plans, four of which had been completed by 1976. It was in 1976 that a revised approach to defense planning and force acquisition was promulgated within the framework of a new document, the *Outline of the National Defense Program*. The 1977 *Defense White Paper* has described the reasoning behind the new approach to defense preparedness:

> Previously, four "defense buildup plans" had been programmed and carried out in order to achieve quantitative and

qualitative improvements in the Ground, Maritime, and Air Self-Defense Forces. The fundamental ideas and principles behind these plans are vague, however, and these plans mainly concerned numerical increases in tanks, ships, aircraft, etc., during the life of each plan. As a result, this series of buildup plans elicited some criticism as mere instruments of equipment acquisition. Calls arose for clarification of the precise ideas and principles—i.e. the basic defense policy—which formed the premise for such buildup effort. In response, the new National Defense Program Outline aims at specific government definition of such basic defense aims as the need to maintain defense capabilities, overall defense posture and the organization of each Self-Defense Force branch.[46]

Thus, the program outline defines the specific defense missions assigned to the Self-Defense Forces. At the same time, it aims at developing a defense structure with specific goals attainable within the foreseeable future. The new plan envisages an outlay of 12.6 trillion yen, or approximately $60 billion at current prices, for the next phase of Japan's defense procurement. And, unlike previous buildups, the new plan avoids a fixed time period for implementation, relying instead on a "rolling budget system." The major consequence of these changes, as a long-time participant in Japan's defense affairs has described it, will be to seek "the goal of a desirable level of force in place of the former procurement of arms."[47] And, although a five-year buildup plan has been retained as an undeclared policy, the ability to react much more rapidly to sudden changes in Japan's strategic situation will put defense planning on a very different basis.

Thus, as indicated throughout this paper, the defense establishment believes that a fundamental change in the Japanese security system will be required when U.S. ground troops withdraw from Korea. For the present, the Defense Agency intends to place special emphasis on strengthening the defense structure in western Japan, centering on Kyushu. Various proposals include the expansion of the ground forces division at Kumamoto, increases in naval forces near the Tsushima Straits, and possible improvement of the facilities at the Tsuiki Air Base. Heretofore, less importance had been attached to the defense of this area than to that of Hokkaido and other northern parts of the country. Thus, concern about the possible effects of renewed conflict in Korea has already begun to affect Japanese defense planning.[48] Whether renewed warfare is initiated by the North or the South, another Korean war would very adversely affect Japanese interests in Northeast Asia. In the context of the U.S. withdrawal, this is not a question that Japan can afford to ignore.

The Korean peninsula also has a strategically vital role in watching the Soviet Pacific Fleet and in helping contain it in the Sea of Japan. Japan would be in great danger and the strategic balance in the area would change markedly if the Soviet Fleet could freely pass through the Korean Straits. The Soviet Union has long desired a year-round port and a military supply sea base outside of the Sea of Japan.[49] The narrow, twenty-five mile Korean Straits are presently defended and guarded by both South Korea and Japan, each country patrolling its own twelve-mile sea belt. Although security arrangements exist between the two countries, there is no actual means for direct consultation and discussion. The present lack of an effective bilateral framework between the two nations poses a serious strategic problem. Experts in the Japanese naval forces criticize this lack of security arrangements that could easily facilitate Soviet fleets penetrating the straits, should the security situation in Japan and Korea deteriorate. For example, Kenichi Kitamura, a retired admiral, pointed out in a seminar held in Seoul that greater collaboration between the two countries would mean defense against Soviet sea superiority and any potential North Korean aggression.[50]

Another key problem is the possible expansion of the Self-Defense Forces' defense perimeter. Now that the 200-mile sea zone has become an actuality, the question is how Japan will patrol and defend the expanded coastal waters. Mounting concern over control of seabed resources, fishery disputes, and related issues have also emerged. For example, Asao Mihara, former director general of the Defense Agency, has testified in the Diet that Japan could legitimately invoke the right of self-defense within the area of joint seabed development between

Japan and South Korea should Japan be attacked.[51] In addition, Admiral Nakamura, chief of staff of the Maritime Self-Defense Forces, has stated at a press conference that the Self-Defense Forces want new policies that reform the Self-Defense Law and the right to regulate foreign ships coming into Japanese sea territory.

A related problem concerns the increased capabilities of new Japanese military equipment. Thus, the proposed introduction of the F-15 fighter plane into Japanese inventories during the early 1980s would lead to a deviation away from a purely self-defense concept, given the plane's range and capacity for midair refueling. Recent debate on this issue reflects yet another aspect in the altered attitude in Japan toward defense issues.

Additional evidence of change can be seen in the attitude of some members of the National Defense Council who feel that Japan should take on more of the defense burden that is still so heavily assumed by the United States. Some council members have presented this view quite strongly, but serious discussions with the U.S. government have not proved possible so long as the present Japanese government lacks a decisive attitude on this question. Indeed, one of the council members has defined equal partnership between the United States and Japan as a relationship in which each shares the same risks. It is obvious that such a relationship does not yet exist.

Discussions are also now underway to transform the National Defense Council into the National Security Council, which would discuss the entire spectrum of Asian issues in terms of Japanese diplomacy and security. This new structure would provide a top-level forum for the discussion of strategic intelligence where formerly it allowed only for discussion of defense buildup plans on a rather businesslike basis. Nevertheless, the concept of such a forum will take time to put into effect. The hope of defense circles is that such a body would help diminish Japan's "military allergy," thus allowing more open debate on the whole range of security issues affecting Japan.

In an overall sense, therefore, a new attitude toward Japan's security has begun to develop. Media debate and discussion are presently raising questions central to defining the nature of Japanese defense policy. The Japanese defense establishment, for its part, wants the ends and means of Japan's military capabilities to be clarified and crystallized far more explicitly. They feel Japan should now prepare to cope adequately with any future problems and changes. Such an attitude in turn spurs expansion of Japan's own military potential and a self-reliant military buildup. Our discussion will therefore now focus on the rearmament issue, with an eye toward future developments.

Japan's Self-Defense Forces: Buildup and Potential

Though numerically small at a total complement of 263,000 men, Japan's Self-Defense Forces are now evaluated by knowledgeable observers as among Asia's strongest military forces. The SDF began modestly in 1950 as a reserve Police Corps of 75,000 men. In 1952, they were converted into a National Security Corps with ground and naval forces. Two years later, they were reorganized into the Self-Defense Forces comprising ground, naval, and air arms. Through a gradual process of expansion and modernization, the SDF are far from insignificant as a combat force. The Ground Self-Defense Forces (GSDF) (composed of 180,000 men, 820 tanks, 350 aircraft, and 190 missiles) already possess greater firepower than the entire Imperial Japanese Army of World War II. The Air Self-Defense Forces (ASDF) comprise 43,000 men, 450 combat fighters, 300 trainers, and 33 transports. The naval arm (Maritime Self-Defense Forces, or MSDF) accounts for 40,000 personnel, 15 submarines, 15 frigates, 30 destroyers, 402 other craft, and 151 aircraft.[52] In 1978, Japan will allocate, at current prices, approximately $10 billion for national defense, placing it seventh among the world's military spenders. In operational terms, Japan during the 1980s will clearly arrive among the front rank of the world's conventional military powers, irrespective of any internal debates over the legitimacy of the SDF.

Japanese defense assessments, plans, and activities have evolved slowly but deliberately as a concomitant of the country's special postwar political environment. The gradual nature of change reflects the inevitable constraint of Japan's early postwar years and the increasing flexibility apparent as the country rebuilt its economic, industrial, and political structures. Table 5.1 illustrates the process of growth in Japan's defense expenditures and Self-Defense Forces.

As is well known, Japan now possesses the second largest gross national product among the non-Communist nations, the third highest in the world. As a result, it has been able since 1960 to maintain defense expenditures continuously below 1 percent of GNP while being able to steadily increase annual defense outlays to their present impressive levels. Indeed, it is an ironic commentary on Japan's changed environment that U.S. leaders and defense analysts are increasingly accusing Japan of having a "free ride" in security matters. Since the early seventies, the United States has pressed Japan to increase its defense spending. Especially in light of Japan's huge trade surpluses with the United States, U.S. officials are particularly concerned with persuading Japan to purchase more U.S. arms, and it appears that the Japanese are now responding to such requests.[53]

Apart from defense purchases abroad, there are increasing indications that political, business, and government groups in Japan now favor the domestic production of a substantial proportion of the country's future weapons needs.[54] Given Japan's technological and industrial sophistication, the local production of arms would appear natural once plans for substantial defense and weapons modernization are put into effect. Japanese firms such as Mitsubishi, Kawasaki, and Tokyo Shibaura have already been engaged in the licensed production of aircraft, missiles, and ground and naval equipment for the SDF, and the same procedure is to apply to future arms purchases from abroad.

However, in order to place an indigenous arms industry on a sound technical and competitive footing, an arms market larger than Japan may be needed. Industry leaders have therefore urged the government to enter the arms export market where customers for Japanese-made weapons already appear to exist.[55] The Japanese government has not made any decision in the matter, but it is apparent that the country's vast industrial

Table 5.1. Japan's Defense Expenditures and
the Strength of the Self-Defense Forces

Fiscal Year	Defense Expenditures (million U.S. $)*	GNP (billion U.S. $)*	Defense Expenditures as percentage of GNP	Total Strength of SDF (thousands)		Remarks
				Authorized	Actual	
1954	375	21.7	1.73	152	146	SDF established
1955	375	24.6	1.52	180	178	
1956	397	27.6	1.44	197	188	All U.S. ground combat troops withdrawn
1957	399	31.2	1.28	214	211	
1958	412	32.7	1.26	222	214	First buildup plan
1959	432	37.8	1.14	231	215	
1960	444	45.0	0.99	231	206	Revision of Security Treaty
1961	510	55.1	0.92	242	209	
1962	594	60.2	0.99	244	216	Second buildup plan
1963	688	71.0	0.97	244	213	
1964	780	82.0	0.95	246	216	
1965	846	90.7	0.94	246	226	
1966	959	105.9	0.91	246	227	
1967	1,075	124.3	0.86	250	231	Third buildup plan
1968	1,172	146.6	0.80	250	235	
1969	1,375	174.2	0.79	258	236	
1970	1,640	203.4	0.81	259	236	
1971	2,252	255.3	0.88	259	234	
1972	2,601	285.7	0.88	259	233	Reversion of Okinawa
1973	3,118	366.0	0.85	260	233	Fourth buildup plan
1974	3,643	438.3	0.83	260	237	
1975	4,424	528.3	0.84	260	238	
1976	5,041	560.3	0.90	260	233	

*Conversion Rate: Up to 1970 $1.00 = 360 yen
After 1970 $1.00 = 300 yen

Source: Seiichiro Ohnishi, "A Recollection and Perspective of the Buildup of Japan's Defense Capabilities," (paper delivered at a seminar of the Harvard University Council on East Asian Studies, Cambridge, Mass., October 11, 1977), p. 31.

base could be quickly geared to armament production. By 1980, Japan will be mass-producing its own air-to-ship missiles, with a sophisticated guidance system, to be fitted to its locally made F-1 fighter plane. According to one Japanese authority, "Though its rocket program still depends on U.S. technology, Japan will eventually be able to produce the propulsion, guidance, and reentry vehicle for a strategic weapons system."[56]

In fact, the civilian space program is quite sophisticated even at the present time. Japanese defense specialists acknowledge privately that the four-stage Mu-series rockets, developed by Tokyo University's Institute of Space and Aeronautical Studies (but built by Nissan Motors, Ltd.), carry specifications and performance characteristics approximating those of the earliest generation U.S. Minuteman missiles. The firm of Mitsubishi is separately developing, under the aegis of the National Space Development Agency, the N-series of booster vehicles whose N-3 version (also called the H vehicle) to be completed by 1985, will be able to lift 1,100-pound payloads into orbit. According to a recent report, "Japan has scheduled the launch of four scientific and eight applications satellites over the next five years and during this time will make a fundamental decision whether to develop advanced launch vehicles or to depend on the United States for opportunities in the space shuttle."[57]

Japan has both signed and ratified the Nuclear Nonproliferation Treaty and there exists no basis other than speculation to suggest that the country plans to develop nuclear weapons. The country obviously possesses such a potential: a large-scale civilian nuclear power program, a plutonium reprocessing plant at Tokai-mura, a pilot uranium-enrichment facility employing the centrifuge process, and adequate nuclear-industrial engineering experience. There has even been some abstract discussion in recent years as to whether Article IX of the Japanese Constitution and its "no war" clause conclusively prohibit Japan from acquiring nuclear weapons. Japan's "nuclear allergy," however, remains extremely strong. This discussion of nuclear and space programs is more a reminder of national capabilities than of intent. In addition, Japan's effort to establish a complete nuclear fuel cycle within the country concerns Japan's interest in achieving national self-sufficiency in energy. It in no way suggests an interest in acquiring nuclear weapons. Should either South Korea or Taiwan detonate a nuclear device, however, the impact on Japan would be phenomenal.

For the present, the potential for major change in Japan's capabilities concerns the acquisition of a far more forceful conventional defense posture. Arguments calling for aircraft carriers and the production of a new series of destroyers and frigates for the MSDF have recently been aired. And, although the official strength of the ASDF is 450 combat fighters, adding the aircraft in the GSDF and MSDF inventories brings Japan's total aircraft strength to 1,250—the fourth largest in the world. Finally, it should be noted that almost half the total strength of the GSDF consists of noncommissioned officers (72,752), and another sixth (23,739) are commissioned officers. This leaves an actual infantry force of only 56,217.[58] The GSDF has, in other words, proportionately too many officers. It is a reasonable conclusion that such a personnel structure is meant to ensure the rapid expansion of infantry divisions should the need ever arise.[59] Thus, notwithstanding constitutional restraints and ambivalent popular attitudes toward military strength, Japan's armed might must be increasingly considered an important factor in the Asian military balance.

Toward the 1980s

The history of the past several decades exerts a powerful hold on Japanese decision makers. Japan has tried to avoid a political role among the big powers, pursuing instead economic growth, and finally becoming an economic giant. Paradoxically, this economic growth has given Japan more flexibility in its strategic planning than it had when it was heavily dependent on U.S. power. Japanese diplomacy, long the stepchild of U.S. foreign policy, has begun to strike out vigorously on its own, a change that challenges Japanese ingenuity and negotiating prowess. Leaders are trying to stop her single-minded pursuit of economic prosperity and rethink her political role in Asia.

Unfortunately, as Osamu Kaihara and Leslie Brown both

point out, the study of security affairs has not been a respectable pursuit in Japan and, consequently, has had little legitimacy within the process of government policymaking. Thus, it is now time for Japan to seriously consider the significance of the strategic changes that the nation is now experiencing.[60] It is still difficult to state what actual political or military role Japan might actually play. Unlike the past, however, Japan's leaders cannot simply wait to react to various international changes. Increasingly, Japan's political and military elite recognize their own nation's strength and its capacity for meaningful, independent action. There seems little doubt that the 1980s will test their ability both to adapt to changing international circumstances and to move Japan even more strongly in autonomous political and military directions.

Notes

1. For important discussions, see Morton Abramowitz, *Moving the Glacier: The Two Koreas and the Powers*, Adelphi paper no. 80 (London: International Institute for Strategic Studies, 1971); Ralph N. Clough, *Deterrence and Defense in Korea* (Washington: The Brookings Institution, 1976); Henry S. Rowen, "Japan and the Future Balance in Asia," *Orbis* 21 (1977):191-210; Kiichi Saeki, "Sonotoki Nihon wa Dosuru" [What Shall Japan Do at That Moment?], *Asahi Journal*, June 17, 1977, 11-15; Fuji Kamiya, "Gaikoseisaku to Shiteno Zaikanbeigun" [U.S. Troops in South Korea as Foreign Policy], *Shokun*, July 1977, pp. 36-57; Donald S. Zagoria, "Why We Still Can't Leave Korea," *New York Times Magazine*, October 2, 1977; and "The U.S. Troop Withdrawal from South Korea and Japan's Security," *Kokubu* (Tokyo: Asagumo Shinbunsha, 1978).

2. Wireless bulletin, Press Office, American embassy, Tokyo, United States Information Service, June 24, 1976.

3. December 12, 1977, interview with Makoto Momoi, Professor, Japan National Defense College. Momoi pointed out that the first official announcement of the planned U.S. troop withdrawal to the Japanese government was made by Vice-President Mondale during a visit to Japan, a week after President Carter took office. According to his account, Mondale explained the plan "in a unilateral way," without any close consultations with Japanese government officials and

without informing the South Korean government in advance.

4. Zagoria, for example, has noted that: "There had clearly been no time for a review of this crucial decision by the Joint Chiefs of Staff, the U.S. military in Korea, the State Department or the president's own top security advisers." Zagoria, "Why We Still Can't Leave Korea."

5. Warnke made these remarks in a meeting with the Japanese press in Washington, D.C. *Yomiuri*, December 21, 1976.

6. *Foreign Policy* 25 (Winter 1976-1977):79-80.

7. See on this point, "Zaikanbeigun to Boeicho" [The U.S. Ground Troops Withdrawal from South Korea and Japan's Defense Agency], *Sekai*, no. 374 (Tokyo: Iwanami Shoten, January 1977); and "Boeicho ni Fukamaru Kikikan" [Deepening Crisis in the Defense Agency], *Yomiuri*, December 8, 1977, p. 35.

8. For further details, see Richard Burt, "U.S. Analysis Doubts There Can Be Victor in Major Atomic War," *New York Times*, January 6, 1978.

9. Interview with Les Aspin, March 9, 1977.

10. Yonosuke Nagai, in *Chuo Kuron*, June 1975, pp. 74-93.

11. *New York Times*, February 25, 1977.

12. Henry Owen and Charles Schultze, eds., *Setting National Priorities: The Next Ten Years* (Washington: The Brookings Institution, 1976), p. 32.

13. As cited in *Defense of Japan* (Tokyo: Japan Defense Agency, 1976), pp. 22-23.

14. *New York Times*, December 13, 1977.

15. Glenn H. Snyder, "Deterrence by Denial and Punishment" in Davis B. Bobrow, ed., *Components of Defense Policy* (Chicago: Rand McNally & Company, 1965), pp. 209-37.

16. Interview with Momoi.

17. *Asahi*, August 7, 1975. In 1975 Prime Minister Miki, pointing out at a Washington press conference that the distance between Pusan and Tsushima is only thirty miles, stated: "Can we say that Korea's peace and security are irrelevant to those of Japan? More than one million regular army troops confront each other across a narrow Demilitarized Zone, spanning the Korean Peninsula which is separated from Japan by only a thin stretch of water."

18. *Asahi*, October 9, 1976.

19. *Defense of Japan*, p. 11.

20. *Asahi*, November 28, 1977.

21. Since this chapter is not an explicit effort to examine the

economic aspect underlying the present circumstances, I have only touched on the implication of present economic issues between two countries—Japan and the United States—and their ramifications for Japan's security policy.

22. *Newsweek*, December 5, 1977. The surplus is large—some $8.5 billion in 1977, part of Japan's global trade surplus of $15 billion. The surplus in 1972 was $4.4 billion. In each of the next three years it was roughly $1.5 billion; then it climbed to $5.3 billion in 1976.

23. *Asahi*, November 11, 1977.

24. See, for example, Andrew H. Malcolm, "Japan's Trade Policy is Self-Protective," *New York Times*, December 11, 1977.

25. "Gaiko Gannen" [The First Year of Japan's Diplomacy], *Asahi*, July 21-30, 1977. See also Yukio Matusyama, "Henkaku e no Shuppatsu" [Advance Toward Change], *Asahi*, January 1, 1978, p. 1.

26. See Masataka Kosaka, *Options for Japan's Foreign Policy*, Adelphi Paper no. 97 (London: International Institute for Strategic Studies, 1973), p. 1.

27. *New York Times*, October 31, 1977.

28. For example, see Leslie Brown, *American Security Policy in Asia*, Adelphi Paper no. 132 (London: International Institute for Strategic Studies, 1977), p. 24.

29. Jerome Alan Cohen, "Asia Senryaku Naki Cartergaiko" [Carter's Foreign Policy Without Asian Strategy], *Asahi*, December 23-30, 1977.

30. I have made this argument in "Beikoku no Shinkyokuto Senryaku" [A New U.S. Far Eastern Strategy], *Kanagawa Shinbun*, January 1, 1978.

31. *Asahi*, August 18, 1977.

32. See the interview in *Newsweek*, January 27, 1975.

33. *Asahi*, September 15, 1977. See also *Asahi*, December 15, 1977.

34. Yonosuke Nagai, *Takyokuka Jidai no Senryaku*, vol. 2, p. 198.

35. *Annual Defense Department Report, FY 1977* (Washington, D.C.: Defense Department, 1976), Appendix, pp. A-19.

36. The number of Japanese Defense Agency officials and Self-Defense Forces officers the Chinese government has invited is quite noteworthy: Iwashima, Osamu Kaihara (former secretary of National Defense Council), Motoharu Hori (former Japanese Navy commander), Kenjiro Mitsuoka (former Ground Self-Defense Force general), and Minoru Maeda (former Japanese Navy vice-admiral).

37. Bernard Weinraub, "U.S. Aides Split on Defense Technology for China," *New York Times*, January 4, 1978.

38. The specific figures were: 83 percent in favor of maintaining

the SDF; 7 percent favored their abolition; and 10 percent had no opinion. Increases in support were particularly marked among males in their twenties. "Opinion Survey for Analyzing Public Relations," *Defense Bulletin* 1 (Tokyo: Japanese Defense Agency, January 1976).

39. For an overall description of this operation and Soviet naval activity more generally, see *The Defense of Japan*, pp. 19-22.

40. Seiichiro Ohnishi, "A Recollection and Perspective of the Buildup of Japan's Defense Capabilities" (Paper delivered at a seminar of the Harvard University Council on East Asian Studies, Cambridge, Mass., October 11, 1977), p. 28. Ohnishi is the former president of Japan's National Defense College.

41. *Asahi*, February 4, 1976.

42. *Asahi*, November 11, 1975.

43. *Outline of the National Defense Program* (Tokyo: Japan Defense Agency, October 29, 1976).

44. *Asahi*, December 12, 1977.

45. *Asahi*, October 15, 1977.

46. *Defense of Japan*, pp. 47-48.

47. Ohnishi, "A Recollection and Perspective," p. 1.

48. Andrew H. Malcolm, "Defense Weaknesses and U.S. Plan for Korean Pullout Spurs Japanese Debate on Military Requirements," *New York Times*, March 22, 1977.

49. On this point, see Robert M. Slusser's remarks in *The Origins of the Cold War in Asia*, Yonosake Nagai and Akira Iriye, eds. (New York: Columbia University Press, 1977), p. 432.

50. *Asahi*, July 7, 1977.

51. *Asahi*, October 28, 1977.

52. Robert Whymant, "Officially Japan Doesn't Have an Army But This is What It Has for Self-Defense," *The Sunday Times* (London), February 5, 1978, p. 9; and, Colonel Wilfred L. Ebel, "Japan's Developing Army," *National Defense*, September-October, 1977, p. 143.

53. "Japan to Buy U.S. Jets At Cost of $4.5 Billion," *New York Times*, December 29, 1977, p. 4. The order is for 100 F-15 "Eagle" fighter-bombers and 45 P-3C "Orion" antisubmarine and reconnaissance aircraft.

54. Susumu Awanohara, "Once More Into the Breach," *Far Eastern Economic Review*, April 7, 1978, pp. 33-34.

55. Ibid., Middle Eastern and Southeast Asian countries have shown interest in acquiring the Japanese-designed and produced Type-74 armored tank, model 62 machine guns and model 64 automatic rifles, the antitank Kam-3, and ground-to-ground Kam-9

missiles. More significantly, the People's Republic of China has shown increasing interest in buying Japanese arms and weapons technology. See, Henry Scott-Stokes, "China Strengthens Its Ties With Japan," *New York Times*, July 23, 1978, p. 9.

56. Tomohisa Sakanaka, as cited in Whymant, "Officially Japan Doesn't Have an Army."

57. *Aviation Week and Space Technology*, March 13, 1978, p. 73.

58. Ebel, "Japan's Developing Army," p. 144.

59. Osamu Kaihara, *Nihon Boeitaisei no Uchimaku* [Inside the Japanese Military Machine] (Tokyo: Jiji Tsushima sha, 1977), p. 253; Brown, *American Security Policy*, p. 4.

60. Yasuhisa Nakada, "Hidogiteki Gaiko no Chokoku" [On Amoral Diplomacy], *Boei Antenna* 297 (1977):14-19 (Tokyo, Japanese Defense Agency).

UA
830
.M54
1980

CANISIUS COLLEGE LIBRARY
UA830 .M54 1980 c. 1
Military power and p

3 5084 00156 6648

UA 830 .M54 1980 $37.16

Military power and policy in
Asian States

CANISIUS COLLEGE LIBRARY
BUFFALO, N. Y.